THE DOMINANT FOCUS
Electrophysiological Investigations

Neuropsychology
A Series of Special Research Reports
Under the Editorship of A. R. Luria

1972 • THE BRAIN AND HEARING: Hearing Disturbances Associated with Local Brain Lesions
A. V. Baru and T. A. Karaseva

1973 • THE DOMINANT FOCUS: Electrophysiological Investigations
V. S. Rusinov

THE DOMINANT FOCUS
Electrophysiological Investigations

V. S. Rusinov
Institute for Higher Nervous Activity and Neurophysiology
Academy of Sciences of the USSR
Moscow, USSR

Translated from Russian by
Basil Haigh

Translation Editor
Robert W. Doty
Center for Brain Research
Medical Center, University of Rochester
Rochester, New York

 CONSULTANTS BUREAU • NEW YORK-LONDON • 1973

The original Russian text published in Moscow in 1969 by Meditsina Press, has been corrected and revised by the author for this edition. The translation is published under an agreement with Mezhdunarodnaya Kniga, the Soviet book export agency.

DOMINANTA: ELEKTROFIZIOLOGICHESKIE ISSLEDOVANIYA

Vladimir Sergeevich Rusinov

ДОМИНАНТА

Электрофизиологические исследования

Владимир Сергеевич Русинов

Library of Congress Catalog Card Number 72-94828

ISBN 0-306-10887-9

© 1973 Consultants Bureau, New York
A Division of Plenum Publishing Corporation
227 West 17th Street, New York, N.Y. 10011

United Kingdom edition published by Consultants Bureau, London
A Division of Plenum Publishing Company, Ltd.
Davis House (4th Floor), 8 Scrubs Lane, Harlesden, NW10 6SE, London, England

All rights reserved

No part of this publication may be reproduced in any form without written permission from the publisher

Printed in the United States of America

Preface

This book describes an electrophysiological investigation of foci of excitation in the central nervous system of animals and man, foci which become dominant in character.

The dominant focus, or dominant as it is usually called by Russian physiologists, is one of the fundamental processes underlying the activity of the central nervous system. The discovery of the mechanisms of formation of dominant foci is important for a deeper understanding of the activity of the human and animal brain and for identifying the pathophysiological mechanisms in lesions of the nervous system.

The book summarizes the results of many years of investigation, both experimental and clinical, into the nature and properties of foci of excitation arising at different levels of the central nervous system.

A brief survey of the literature on Ukhtomskii's theory of the dominant is given in Chapter I and the present state of this theory is reviewed. The other chapters of the book deal with the results of research in the field of dominant foci undertaken in the author's laboratory. The experimental data relating to changes in the level of the steady cortical potential during the formation of the dominant focus and conditioned reflex are examined. The results obtained by polarization of the cortex with a weak direct current in order to form a dominant focus, and also by polarization of the reticular system of the brain stem and specific and nonspecific nuclei of the thalamus and hypothalamus are given. Electrophysiological studies of the character of excitation in dominant foci, and the relationships between cortical foci of excitation and foci of excitation induced experimentally in various subcortical structures are described. The author submits a working hypothesis on mechanisms of dominant formation based on electrophysiological findings and on conversion of the dominant focus into a "temporary connection" as seen in conditioned reflexes. Much attention is paid to experimental data on excitation traces associated with the dominant focus and their possible significance for mnemonic mechanisms. The results of electrophysiological investigations of foci of excitation in healthy subjects and in patients with organic brain lesions are examined, with particular reference to the dynamics of changes in evoked responses in man. Results obtained by auto- and cross-correlation analysis of the electrical activity of the human and animal brain in the presence of foci of excitation are also examined.

The book is intended for medical scientists, biologists, physiologists, biophysicists, and workers in other specialties interested in neural processes. It could also serve as additional reading for advanced students of medicine and biology.

Contents

Chapter I. Foci of Excitation and Conditions for Their
Conversion into Dominant Foci
    1. A brief historical survey 1
    2. The summation reflex and the dominant 7
    3. The swallowing dominant. Two foci of excitation
       with rhythmic action in the central nervous
       system 9
    4. Disinhibition of the dominant 12
    5. The swallowing dominant and conditioned reflex . . . 13
    6. Importance of the time factor 14
    7. Wenckebach periods in the rhythmic activity of the
       focus of excitation 16
    8. Interaction between a cortical focus of excitation
       and the swallowing center 20
    9. Foci of excitation induced by strychnine in sub-
       cortical structures 24
    10. A focus of excitation in the visual cortex 28
    11. Corticofugal influences from foci of excitation 29

Chapter II. The Problem of Stationary Excitation and Changes
in the Cortical Steady Potential during the Formation of
Dominant Foci and Temporary Connections
    1. Two approaches to the theory of excitation 33
    2. Slow long-lasting potentials in the cortex 34
    3. Interaction of slow potentials with evoked re-
       sponses and dendritic potentials 39
    4. Hypotheses on the genesis of slow electrical po-
       tentials 43

5. Infraslow rhythmic potentials 47
6. The problem of stationary excitation 49
7. Slow potentials in the cortex during defensive conditioning 51
8. Long-lasting depolarization of the single neuron. . 56

Chapter III. A Model of the Cortical Dominant Focus Produced by Weak, Steady Current
1. Polarization of the rabbit motor cortex by steady current 71
2. Conversion of the dominant into inhibition; the double strength optimum of the direct current . . . 78
3. Trace phenomena associated with the dominant focus and memory 91
4. Foci of excitation in the central nervous system evoked by a pulsating current. Pavlygina's experiments 95

Chapter IV. Diffuse Effects in the Central Nervous System. The Reticular Formation and the Dominant Focus
1. The data of classical physiology on diffuse effects in the central nervous system 107
2. The reticular formation of the brain stem and "arousal reactions" 109
3. Similarity between EEG of the motor and visual cortex in the presence of a motor dominant reinforced by photic stimulation 112
4. Electrical activity of the lateral geniculate body and caudate nucleus in the presence of a cortical dominant focus 115
5. The cortical dominant focus in long-term experiments 118
6. Electrical activity of the medial geniculate body and caudate nucleus in the presence of a motor dominant reinforced by acoustic stimulation 120

Chapter V. Cortical-Subcortical Relationships, Thalamo-Cortical Connections, and Dominant Foci in the Hypothalamus
1. Polarization of thalamic areas 125
2. Polarization of the medial nuclei of the thalamus 129

3. Polarization of the hypothalamus 130
4. Polarization of individual layers of the cortex 134
5. The recruiting response. 137
6. Functional connections of the nonspecific thalamo-cortical system with cortical neurons. 142

Chapter VI. Relations between Specific and Nonspecific Thalamo-Cortical Systems. Correlation Analysis of EEG Rhythms
1. Relations between the specific and nonspecific thalamo-cortical systems in man based on the study of evoked potentials. 149
2. Correlation analysis of the EEG of the motor and visual cortex during polarization of the motor cortex 154
3. Foci of increased excitability evoked in the cortex by repetitive stimulation 158
4. Generators of rhythmic electrical activity in the human cortex, particularly the motor cortex, as revealed by correlation analysis 161
5. Correlation analysis of rhythmic driving of the human EEG by flashes. 168
6. The rhythmic driving by flashes when a pathological focus is present in the brain. 173

Chapter VII. Convergence of Impulses on Neurons of the Motor Cortex, and the Motor Dominant
1. Functional convergence and polysensory neurons of the motor cortex. 175
2. Changes in ultrastructure of the motor cortex following polarization by weak direct current. 184
3. General scheme of processes in the central nervous system during anodal polarization of the motor cortex 186

Conclusion 191

Bibliography 205

Chapter 1

Foci of Excitation and Conditions for Their Conversion into Dominant Foci

1. A Brief Historical Survey

The theory of the dominant focus, or dominant as it is generally called in Russian physiology, was created by A. A. Ukhtomskii. The term "dominant" was first used and conceived as a general principle governing the activity of neural centers in 1923, but the experimental material on which Ukhtomskii based his theory was published in 1911.

The dominant is defined as the prevailing reflex in an animal at each moment of its activity. More precisely, the dominant is the temporarily prevailing reflex system which directs the activity of neural centers at a given moment. If the center has been made ready to react by preliminary, weak influences from the appropriate receptive field and if its excitability is raised, excitation will grow in it as the result of impulses reaching the central nervous system quite regardless of the site and modality of stimulation. Prepared beforehand by impulses of its own reflex arc, and reinforced by summation with impulses from any extraneous source, the excitation arising in the dominant center inhibits other reflexes to stimuli usually evoking them.

Under normal conditions the dominant is a functional combination of neural centers consisting of a relatively mobile cortical component and subcortical, autonomic, and humoral component.

The most important experimental results and the conclusions from them which led Ukhtomskii to introduce and develop his concept of the dominant were as follows.

Variations in motor effects arising in response to cortical stimulation depend not only on the strength of that stimulation, but also on intracentral influences from other centers as they are brought into the sphere of response. Lasting and prolonged excitation of any center, especially of one controlling "chain reflexes," is incompatible with stable excitation of other centers connected functionally with it: they are reciprocally inhibited.

Ukhtomskii sought the clue to the variability of the effects of cortical stimulation in the waking animal in central responses taking place in the nervous system at the time of stimulation. The cortical "center," responsible for a certain function takes over other functions when the mechanism which it carries acts as part of a wider central mechanism.

"Normal cortical activity does not take place as if it were based on some constant, static function, fixed once and for all time, of foci as carriers of individual functions," wrote Ukhtomskii, "it is based on a never-ending intracentral dynamics of excitation in cortical, subcortical, medullary, and spinal centers, determined by the variable functional states of all these structures."*

These results were obtained, in particular, by studying interaction between the swallowing reflex evoked by an adequate stimulus (introducing water into the animal's mouth) and movements of the hind limbs in response to electrical stimulation of the corresponding point of the cortex. Inhibition of reflex movements of the hind limbs by excitation of the apparatus of deglutition occurred on the average 30 sec after the beginning of swallowing movements. The inhibition must not be understood to imply that the cortical centers became refractory at that time to electrical stimulation. The stimulated cortical focus for the hind limb continues to respond, but the responses were modified: it now reinforced the excitation in the swallowing center, both when electrical stimulation still maintained typical cortical responses in the hind limbs and also when these responses in the hind limbs were inhibited.

---

*A. A. Ukhtomskii, Collected Works, Vol. 1, Leningrad University Press, Leningrad (1950), pp. 81-162.

1. A BRIEF HISTORICAL SURVEY

That stage of the experiment when the phenomenon of reinforcement of excitation in the swallowing center coincides with inhibition of the corresponding cortical motor responses corresponds, according to Ukhtomskii, to the functional moment which, it must be assumed, is the basis of reflex attention. This link between the responses is easily dissociated in time: the phenomenon of reinforcement of excitation of the swallowing center from the cortical focus of the hind limb persists much longer than the phenomenon of inhibition of the hind limbs by excitation of the swallowing center. Results obtained with the normal and free animal suggested an even closer and regular dependence of cortical motor innervation on intracentral influences. With the spread of excitation beyond the limits of localized responses, the effect of particular cortical foci on particular efferent systems becomes variable in time, and the "center" loses its significance of an apparatus with one particular, special functional role. There is no basis for considering that permanently fixed, static mechanisms exist between the centers. Antagonistic relationships between centers can be determined by temporal conditions, they may arise dynamically. Ukhtomskii considered that there are no grounds for drawing a hard and fast line between the sphere of "reciprocal innervation" and the region of innervation of anatomical antagonists. Other, even highly dissociated centers may form such reciprocal relationships when they regulate a particular state of an organ. The normal role of the center in the organism is not its only quality, static and unchanging, but one of several possible states in which it may exist.

These were the principal findings and conclusions obtained by Ukhtomskii from his investigations into the effects of extraneous central influences on cortical motor responses. Behind the relationship described above, Ukhtomskii saw the important fact of normal activity of the central nervous system, and he imagined that constantly changing tasks in a constantly changing environment momentarily give rise to "dominant foci of excitation" in the central nervous system. He set out from the conviction that the ability to form dominant foci of excitation (dominants) is not an exclusive property of the cortex, but a common property of the centers. The dominant was defined as the general principle governing activity of the central nervous system, and this was subsequently confirmed by Vinogradov (1923) and by Kaplan and Ukhtomskii (1923) in experiments with dominant foci in spinal frogs, and by Katsnel'son and Vladimirskii (1923) in experiments on gastropod mollusk ganglia. Increased excitability in the dominant center can be formed not only by preliminary preparation with adequate stimuli, but also by the action of hormones.

Uflyand (1925) found dominant patterns of behavior in the clasping reflex of spring frogs, an example of a natural hormonal dominant focus. Vetyukov (1930) found that the spinal dominant is best reinforced, not by strong, but by weak repetitive stimuli.

Must the dominant focus of excitation (or dominant, as it can more conveniently be called) be regarded as topographically the sole point of excitation in the central nervous system? A more appropriate concept of the dominant is that of a synchronously acting constellation of centers with the optimum level of stationary excitation for a given response at different levels of the brain and spinal cord and also in the autonomic nervous system.

In the spinal frog a specially sensory and a specially motor dominant can be evoked in centers developing different effects on the same hind-limb reflex.

Beritov (1910) showed that the local application of strychnine to areas of the spinal cord can give rise to a restricted and reasonably stable focus of increased excitability. Ukhtomskii and Kaplan found that these foci in the spinal cord influence spinal responses and that the character of this influence can be varied at will by changing the foci in the spinal cord. These workers produced foci of excitation at a certain level of the dorsal half of the spinal cord by strychnine, or at the same level of the ventral half of the spinal cord by phenol. They used the reflex of rubbing the hind limb as indicator of spinal cord activity.

A typical dominant focus arose in the experiments with strychnine: various kinds of stimuli applied to the animal evoked a reflex movement or reinforced an existing reflex movement. This was a sensory dominant, for after division of the corresponding dorsal roots it could no longer be evoked by strychnine and stimulation of the skin gave rise to ordinary rubbing, aimed at the actual point of stimulation.

Consequently, the sensory dominance was manifested not only as a lowering of the threshold of excitability, but also as a change in the direction of coordination of the reflex: it was directed at the zone of hyperesthesia and not at the point of actual stimulation.

This particular dominant was manifested best in response to weak stimuli. Stronger stimulation obscured the picture, primarily through hastening the onset of strychnine convulsions.

After application of phenol to the ventral half of the spinal cord a substantially different picture was observed: one of the first features was that in response to a wide variety of remote tactile stimuli, the "poisoned" limb initiated the response, but the rubbing reflex was aimed at the point of actual stimulation. This feature, the ability to initiate the reflex, was most stable and persisted for many hours.

In their investigation of the spinal dominant, Vinogradov and Konradi (1928) showed that the dominant can be "symmetrically transferred," when the level of excitation in the focus reaches its optimum and an induced secondary dominant develops at the symmetrically opposite point of the body. By subsequent analysis of dominants of this type, Konradi (1930) showed that they were regularly converted into inhibition.

Airapet'yants and Balakshina (1933), in their experiments on animals with complete transection of the spinal cord, demonstrated the development of reciprocal inhibition in the centers of the hind limbs, in the presence of a cortical dominance, through the autonomic nervous system.

The concept of mobile reciprocity in the dynamics of neural centers led Ukhtomskii to accept the principle of the dominant. A new and regular link between the centers was thus discovered, and in it he saw a more general and dynamically mobile example of the well-known reciprocal innervation previously established in a narrower and more static form for anatomical antagonists (Sherrington, 1911).

Outward manifestations of the dominant are steadily sustained activity or the active posture of an organism, reinforced at a given moment by a variety of stimuli.

The principal features of the dominant are increased excitability, stability of excitation, ability to summate excitation, and inertia, i.e., ability to maintain and continue excitation once it has started, and when the initial stimulus to excitation is over. The ability to summate excitation is particularly important. It is not the strength of excitation in a focus which makes it dominant, but the ability to summate and accumulate excitation from impulses derived from random stimuli of any modality. The state of "preparedness" for a particular response or the "tendency" toward a response, produced by the action of random stimuli, is the expression of the dominant.

Excitation in the dominant focus is initially at such a low level that it cannot be detected until impulses from indifferent stimuli start to undergo summation in it, and demonstrate its dominant role in the current response. Diffuse waves from stimuli of any modality will excite all centers which are sufficiently excitable at that particular moment, but the dominant is formed only in that center which can summate the excitation.

Summation (accumulation) of excitation in the centers is closely connected with the problem of traces, with the physiological role of the time factor. "...We must suppose that in the cell, especially in the nerve cell and even more in the cortical cell, the transmission of traces from one moment to another must play an exceedingly important role."* It is only during the response itself that excitation in the dominant focus rises to high levels.

The physiology of the dominant is attended by several problems: as a factor in behavior the dominant is closely linked with higher nervous activity and with psychology; as a temporarily prevailing focus, under certain conditions evoking dominant responses, whether in the spinal cord, in the mesencephalon, or in the cortex, it is connected with problems in the general physiology of the central nervous system calling for "the most painstaking study of intimate intercentral activity." The point at issue in this last case is the physiological mechanisms of conversion of ordinary foci of excitation into dominant foci.

---

*A. A. Ukhtomskii, Collected Works, Vol. 1, Leningrad University Press, Leningrad (1950), pp. 208-220.

The basic purpose of this monograph is to examine the conditions of conversion of foci of excitation in the central nervous system into dominant foci, to create a model of cortical and subcortical dominants, and to examine a number of problems arising during the investigation of the dominant by electrophysiological methods and, in particular, the role of specific and "nonspecific" neurons in the formation of the motor dominant, the slow, long potential as a reflection of stationary excitation, the functional role of long electrical potentials in the cortex, and their possible genesis.

2. The Summation Reflex and the Dominant

Pavlov (1932) wrote: "I. M. Sechenov's book 'Reflexes of the Brain' contains, in a clear, precise, and fascinating form, the basic idea of what I am now working on. What a great effort of creative thought was required at that time, with the physiological data then available, in order to produce this idea! But once it had been created, the idea grew, matured, and has now become a scientific lever directing the vast amount of work which is being done today in the field of brain research."*

Foci of excitation in the central nervous system are essential to all activity of the organism. Having assumed their dominant character, they exert a decisive influence on the course of responses as they take place. The concept of dominant focus implies Ukhtomskii's "constellation," formed as a system actually in the course of current activity, at all levels of the central nervous system and in its different parts, but with the primary focus in one of its parts and with variation of importance to the functions of individual components of the system. In other words, the individual components of a given constellation may assume a different functional role in conjunction with other constellations, or during participation in other activities.

Pavlov distinguished between the summation reflex or dominant and the conditioned reflex. Ukhtomskii also considered that the dominant is an independent entity. It is only partly true to say that the dominant is a summation reflex. The dominant is based

---

*I. P. Pavlov, Complete Collected Works [in Russian], Vol. 3, Moscow and Leningrad (1951), p. 125.

on a summation reflex, but it is itself more complex. This is clear from the following example. Stimulation of a rabbit's hind limb by an electric current of subthreshold strength in the receptive field of the flexor reflex evokes reflex flexion of that particular limb, not in response to the first stimulus, but to a subsequent stimulus. This reflex is based on the ability of the central nervous system to summate subthreshold stimuli. If, however, a dominant focus is created, for example by decremental repetitive electrodermal stimulation or by a chemical agent, in the center corresponding to one of the forelimbs, and a hind limb is then stimulated, the animal will respond by flexion, not of the hind limb, but of the "dominant" forelimb. This is a reflex of a higher order — a dominant.

The dominant appears on the basis of summation, but its physiological mechanisms are more complex than those of the simple summation reflex. The summation reflex does not possess such high inertia as the dominant. During the summation reflex, if reciprocal inhibition occurs, it does so in anatomically secured connections (for example, centers for the flexors inhibit centers of the extensors at the height of the reflex, and vice versa). Reciprocal relationships during the dominant are not secured anatomically by permanent reflex arcs: they are established during the response itself in accordance with the conditions of current activity.

The physiological mechanisms of conditioning are connected genetically with summation, but a distinction can be drawn between the simple summation reflex and the complex summation reflex constituting the dominant. A conditioned reflex cannot be reduced to a dominant nor to a summation reflex, but both these last two concepts play an important role, by virtue of their mechanisms, in the formation of a temporary connection. A conditioned reflex is formed from a dominant, a dominant is formed from a summation reflex. At the beginning of its formation, the conditioned reflex is analogous to the dominant in its physiological mechanisms, but later it becomes something essentially different. The dominant is formed at all levels of the central nervous system and the presence of the cerebral cortex is not essential for its formation. The cortex in higher animals is essential for conditioning.

The dominant "passes through" a summation reflex initially before it becomes a dominant. The conditioned reflex is initially a dominant before it becomes a conditioned reflex. That is the scheme of the relationship between the physiological mechanisms of the summation reflex, the dominant, and the conditioned reflex. The role of functional mobility (lability) and rhythm binding in dominant formation has been analyzed by Golikov (1950).

Ukhtomskii (1911) conducted his investigations on animals in acute experiments. In my laboratory the view was adopted that chronic experiments could demonstrate new features of the dominant which acute experiments could not reveal because of their very nature.

3. The Swallowing Dominant. Two Foci of Excitation with Rhythmic Action in the Central Nervous System

Kuznetsova (1957) investigated the properties of the swallowing dominant in long-term experiments on rabbits. A special device was made to ensure a uniform supply of water and its precise measurement. The volume of water entering the rabbit's mouth per unit time was strictly specified and was always monitored. During and after introduction of the water, indifferent stimuli were applied. The regular rhythm of swallowing, constancy of the latency and interval between swallows so long as the parameters of administration of water were unchanged, plus the regular changes in the intervals depending on the force of the stream of water, all indicated that the stimulating factor evoking the swallowing reflex in these experiments was the accumulation of a specified volume of water at the back of the mouth.

The swallowing center was sensitive in its rhythmic activity to all types of inadequate stimulation. Information on the functional state of the swallowing center, as it acquired the character of a dominant, was obtained continuously from the rhythms of swallowing.

In experiments conducted in this way two rhythmically functioning foci of excitation were found in the central nervous system: the respiratory center and the swallowing center. The rhythms of their operation are different. The relationships between these two functioning foci of excitation will now be examined.

The first point to note is that the focus of excitation arising in the swallowing center, with a constant rhythm of action, influences the rhythm of operation of the other center: a change in the normal rhythm of respiration takes place in the interval between swallows. After the middle of an interval, the respiratory movements become steadily slower. There is apnea at the actual moment of swallowing, and immediately thereafter the rhythm of respiration is restored. The results clearly demonstrate the inhibitory effect of the swallowing center on respiration.

Inhibition of one center by the other was not the only form of interaction observed between the two rhythmically functioning foci of swallowing and respiration. The reverse effect was also shown to exist: that of the respiratory center on the swallowing center. The swallowing center, with a slow rhythm of function, begins to follow the rhythm of the respiratory center, but only under certain conditions of interaction between the centers, when the dominant focus inhibits the respiratory center and the latter begins to function at a slower rhythm. At that moment waves in rhythm with the respiratory movements are distinguishable on the swallowing curve. These waves appear soon after the introduction of water, just preceding each swallowing movement, and they disappear after swallowing.

In general the rhythm of the respiratory center is not assimilated, and it is only slightly slowed. The basic conditions for any assimilation of this rhythm is the preliminary appearance of stationary excitation in the swallowing center, as reflected by an increase in tone of the muscles of deglutition.

In other words, when the rhythmically acting swallowing center begins to acquire the properties of a dominant focus and to produce reciprocal inhibition of the other rhythmically acting center, a focus of stationary excitation is formed in the dominant center itself. A gradual increase in tone of the muscles of deglutition at this time is the peripheral reflection of the stationary excitation of the swallowing center.

Let us examine, using the swallowing center as the example, how extraneous stimuli influence the rhythmically functioning dominant focus.

3. THE SWALLOWING DOMINANT

In G. D. Kuznetsova's experiments extraneous stimuli were applied at various times between swallows. The frequency of swallowing was increased if the stimulus was applied 5-10 sec after the previous swallow, i.e., in the first half of the interval between two swallowing movements. If the stimulus was applied later, in the second half of the interval between swallows, inhibition of swallowing was observed or the rhythm of swallowing was unchanged. If, on the other hand, an increase in the frequency of swallowing was observed, as a rule it was not the interval during which the extraneous stimulus was applied that was shortened, but the next interval.

When indifferent stimuli of varied strengths were applied during a series of swallows, the changes in swallowing were found to depend on the strength of the stimuli applied. Shortening of the interval between swallows was observed when weaker stimuli were applied. Application of strong stimuli as a rule increased the interval. If indifferent stimuli (acoustic, photic, tactile) were applied during the aftereffect of swallowing, when the introduction of water had ceased, swallowing took place in response to these stimuli. Similar responses to indifferent stimuli were observed during the first 60 sec after stopping the introduction of water, but they could still be observed sufficiently constantly even after 2 min. Application of the stimuli at longer intervals evoked swallowing only rarely. If swallowing did not take place in response to a single stimulus, it could be obtained by repeating that stimulus several times in succession: initially only the tone of the muscles of deglutition was slightly increased in response to the weak stimulus, but complete swallows followed its repeated application.

In his writings Ukhtomskii states that the dominant focus in the swallowing center is formed gradually, after a series of strong swallowing movements. This was also observed in Kuznetsova's experiments, in which during application of extraneous stimuli swallowing was not reinforced until the second minute of introduction of water (after 3 or 4 swallows), and in experiments in which small volumes of water were given it was reinforced 3-4 min after the beginning of the infusion. From this point of view, the results obtained by Ukhtomskii and Kuznetsova are identical. However, there are differences, caused by the fact that Ukhtomskii conducted his investigations in acute experiments on animals which had undergone a major operation and had only just recovered from the anesthetic. In his experiments water was injected into the animal's (cat's) mouth in a jet. The swallowing movements usually followed each other at intervals of 3 sec. Consequently, in Ukhtomskii's experiments the excitation of the swallowing center was stronger than in those of Kuznetsova. Dominant properties are known to be more marked during a low frequency of swallowing arising in response to the introduction of a very small quantity of water.

4. Disinhibition of the Dominant

Disinhibition of foci of excitation which have gone into a state of inhibition in peripheral nerves is a characteristic and well-known phenomenon (Vinogradov, 1914-1915; Vasil'ev, 1924; Vorontsov, 1925; Rusinov, 1930b, 1934; Petrov and Lapitskii, 1926, and others). Can a dominant focus be disinhibited? In Ukhtomskii's experiments "on a very tired animal" toward the end of an experiment which had lasted for many hours, when introduction of water into the mouth no longer evoked a swallowing reflex, and cortical stimulation evoked weak and inconstant responses in the muscles, electrical stimulation of the motor cortex against the background of administration of water still continued to induce swallowing. Ukhtomskii also attributed this phenomenon to the summation of excitation. In other words, every case of function of the dominant focus in response to extraneous stimuli he regarded as the result of summation of excitation in the focus.

In Kuznetsova's long-term experiments in which stimuli were applied against the background of inhibited swallowing, when the frequency of the swallowing movements was less than would be expected from the quantity of water introduced into the animal's mouth, extraneous stimuli caused an increase in the frequency of swallowing over the background level, i.e., disinhibition occurred.

The following conclusion can be drawn from these results: to the four well-known features of the dominant (section 1) another must be added — ability of the dominant focus to be disinhibited.

The conclusion that the dominant focus can be disinhibited is important not only to a fuller understanding of the properties of the dominant, but also to an understanding of the physiological mechanisms of the formation of temporary connections, especially in the initial period of conditioning.

5. The Swallowing Dominant and Conditioned Reflex

When the mechanism is analyzed whereby swallowing appears in response to acoustic stimuli after cessation of administration of water, the possibility must be considered that a conditioned reflex may have been formed to these stimuli during their previous application in conjunction with the administration of water.

Kuznetsova investigated the relationship between the swallowing dominant and conditioned reflexes (CRs) and showed that during the ordinary method of conditioning, in which the conditional stimulus (CS) preceded swallowing, a reflex was formed at the 10th-15th combination. At the beginning of conditioning, induction was observed between the center receiving the CS and the center of the unconditioned reflex. Later, with an increase in the number of combinations, the level of excitability increased in the unconditioned center, the latency to the appearance of swallowing was shortened, and summation of excitation was clearly observed in the center of the unconditioned reflex (UR). Judging from the results of this experiment, the principal mechanisms of formation of the temporary connections in the initial period of conditioning are induction relationships (external inhibition) and summation in the center for the unconditional stimulus (US). The center of unconditional stimulation changes its functional state during the formation of the CR and becomes a focus of summation.

The conditioned swallowing reflex was formed in response to various stimuli both in the usual way and by "overlapping." An interesting difference was found. If the CR was formed by "overlapping," i.e., by applying an indifferent stimulus during development of the swallowing dominant, in response to the first combinations, summation takes place: swallowing movements become more frequent and are not inhibited, as during conditioning in the ordinary way. With the "overlapping" method there are no induction relationships. The CS immediately begins to act under the control of the swallowing dominant.

"Overlapping" is considered to interfere with CR formation. Kuznetsova's (1957) experiments to study the formation of conditioned swallowing reflexes by "overlapping" showed that these difficulties arise because the time factor in intervals several seconds in duration is not always taken into consideration. The effect of the CS was opposite depending on whether it was applied in the first (5-10 sec) or second half of the interval between swallowing movements.

6. Importance of the Time Factor

Vvedenskii introduced the time factor into physiology as a parameter determining the course of an ongoing response. The different forms in which this parameter was determined in his first investigations were finally expressed in the law of relative functional mobility (lability).

Microintervals of time determine the fate of excitation waves in peripheral nerves. The course of a response is even more clearly dependent on the temporal factor in the central nervous system, and in its higher levels. The importance of this factor is seen particularly clearly during formation of CRs. For instance, in a delayed CR, the CS, continuing for a considerable time, remains the same but the response during different periods of its action becomes clearly different: it is only at times close to addition of the UR that the stimulus acquires conditional significance (Pavlov, 1932). The role of a time interval measured in milliseconds in the mechanism of CR formation was demonstrated by Asratyan (1960) and his collaborators Pakovich (1960) and Pressman (1960). Pakovich showed that if an acoustic stimulus of moderate intensity is combined with electrical stimulation of the skin of a dog's limb, no motor reflex is produced if the acoustic stimulus precedes the electrodermal by 100 msec or less. A reflex is formed only if the interval between the beginning of acoustic and the beginning of electrodermal stimulation is longer than 100 msec. Parameters such as the latency, strength, duration, and stability exhibit their dependence on the temporal factor particularly clearly with intervals of 100-500 msec.

6. IMPORTANCE OF THE TIME FACTOR

The very strong defensive reflex in dogs is inhibited if the interval between the beginning of the CS and US is less than 100 msec. The inhibited reflex reappears if the interval exceeds 100 msec, and its recovery is proportional, within certain limits, to the excess over this threshold time interval.

Pressman showed that interaction between differential and positive motor CRs in cases in which short conditional stimuli are applied successively at microintervals of time, follows a different course depending on whether the stimuli are addressed to the same or to different sensory systems.

These results indicate the role of temporal factors in determining the course of neural processes. In other words, the functional state of structures of the "working constellation," modified by the influence of stimuli, determines, other conditions being equal, the outcome of the response to the stimulus.

It is interesting to examine the properties of a focus of excitation before it becomes dominant. What is the character of excitation in the focus — is it stationary or rhythmic? If it is rhythmic, can the rhythm be preserved when the focus is in an inhibited state? Can a cortical focus of excitation transmit its assimilated rhythm to subcortical structures, to the reticular system? Do foci of excitation in the central nervous system share common properties with foci formed in other systems of the body?

These and other questions arising during the investigation of foci of excitation are important to the understanding of general problems in neurophysiology, for the dominant is only one stage in the functional evolution of the focus of excitation in the central nervous system: not every focus of excitation becomes dominant. Moreover, foci of excitation, especially those functioning rhythmically, exist not only in the central nervous system.

The dominant is one form of activity of foci of excitation in which the focus is established at a certain level of stationary excitation enabling the summation of excitation reaching the nervous system to take place, and establishing the rhythm of activity at the optimum level for the prevailing conditions, at which this focus responds predominantly to the external stimulus while, at the same time, other functioning foci are inhibited.

7. Wenckebach Periods in the Rhythmic Activity of the Focus of Excitation

Let us examine some of the questions concerned with the general principles of function of foci of excitation mentioned above.

In an animal in the resting state, in the absence of extraneous stimuli, the rhythm of swallowing was uniform provided that the conditions of introduction of water into the rabbit's mouth remained unchanged. The latent periods in a series of swallows and the intervals between them were inversely proportional to the volumes of water entering the animal's mouth per unit of time. The regular rhythm of swallowing usually observed in water-administration experiments was disturbed if the rabbit had been deprived of water on the previous day, and also in some experiments carried out in summer in hot weather, i.e., when the level of excitation of the drinking center was increased. In such cases the rhythm of the swallowing center was modified. One type of change attracted attention because these changes took place in accordance with a definite rule which had hitherto been known only for changes in the heart rate, changes known in the literature as Wenckebach periods (Wenckebach and Winterberg, 1927). According to this rule, there is a gradually progressive slowing of conduction from the atria to the ventricles, reflected in the electrocardiogram (EKG) by a gradual shortening of the P–Q interval. When the decrease in conductance reaches a certain level, one of the ventricular contractions is omitted, for the impulse from the atria is not transmitted to the ventricles. Immediately after this phenomenon the Wenckebach period starts again. The number of cardiac contractions in each period varies in different patients. Similar periods have been observed in the heart of cold-blooded animals by Samoilov (1929) after the creation of a "bridge" by partial division of the ventricle, and by Borisov and Rusinov (1940, 1949) in the presence of a similar "bridge" and also by the action of KCl on the frog's heart.

Essentially the same phenomenon was exhibited in the rhythmically functioning swallowing center of the rabbit under certain experimental conditions. After application of the water stimulus, the double swallowing movements were initially very close together, but later the second swallowing movement began to move grad-

7. WENCKEBACH PERIODS

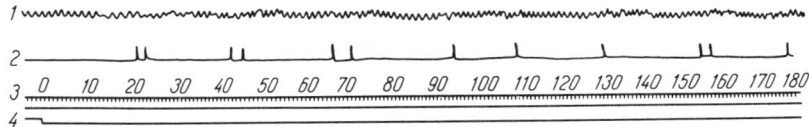

Fig. 1. Wenckebach phenomenon in the activity of the rabbit's swallowing center (after Kuznetsova): 1) respiration; 2) swallowing movements; 3) time in seconds; 4) marker of stimulation: step down shows time of introduction of 2.8 ml water into rabbit's mouth.

ually further away, and eventually it disappeared (Fig. 1). After the single swallowing movement, double movements again were observed and the period started again.

The gradual increase in the latency of the second "spike" and, ultimately, its complete disappearance, with subsequent repetition of the whole period is noteworthy for two reasons: as a phenomenon characteristic of rhythmically functioning foci in general, and as a phenomenon in which the inhibitory response is found in a simpler form as a series of increasingly inhibited impulses.

As the experiments on the comparatively simple preparation with the "bridge" of heart muscle showed, the gradual increase in duration of the R^1-R^2 intervals (from the base to the apex of the ventricle) in these periods is connected with a change in the functional state of the "bridge" itself, and with a gradual decrease in its conductance. The gradual increase in duration of the R^1-R^2 interval in the period is evidence that each passing impulse is a factor actively influencing the "bridge." Usually the wave of excitation, as it travels from the base to the apex of the ventricle, has no such effect. If, however, the functional state of the intermediate link along which the wave of excitation must pass in its course is strongly depressed, each wave will be increasingly delayed as it passes through the "bridge," and this would be reflected in a gradual increase in length of the interval.

Disappearance of one of the successive series of impulses is explained by the following rule. If the lability of the intermediate focus is reduced to such a degree that the wave of excitation is prolonged during its transmission, consequently all the conditions will be present in that focus for the wave of excitation to be converted for a certain time, during the passage of one impulse in the series, into a stationary state, i.e., to be inhibited. The next wave, arriving at the "bridge" against this background of strongly reduced lability, reinforcing its stationary excitation, thus apparently obstructs its own onward passage.

How can the renewal of the period be explained? It is a perfectly regular phenomenon of afterexaltations following inhibition, a "rebound," emphasizing that omission of the apex beat of the heart was in fact due to inhibition.

A similar pattern in the form of Wenckebach periods has been found during the passage of an excitation from the right to the left half of the frog's ventricle when, as a result of application of KCl to the sinus venosus, a division was apparently created between the right and left ventricles, and also during the conversion of local excitation into a spreading wave in the ventricle (an R wave) (Borisova and Rusinov). In the second case the latency, i.e., the time from the beginning of the local ventricular potential to the moment of its conversion into an R wave, increases with each successive contraction of the heart until one beat disappears, and only the local potential is recorded on the EKG. These periods repeat several times and then disappear.

What is implied by this gradual increase in latency of the spreading waves against the background of the local potential during each successive excitation of the ventricle?

First, that there is a certain type of transient block at the place of origin of the spreading wave. If there were no such area of blocking at the point where the local excitation is converted into an impulse, there would be no grounds for the gradual increase in time required for the local activity to be converted into a spreading wave of excitation.

Second, each impulse generated in this particular case augments the local block still further, i.e., the impulse, by its appearance, actively influences the functional state of the blocking zone.

Third, the resulting block, when only the potential of local excitation is recorded on the EKG as a monophasic wave, is in fact an inhibitory state. It is this which explains why a rebound exaltation takes place at the next excitation, in the form of the more rapid appearance of an R wave and a decrease in its latency.

According to all the evidence, the inhibition obtained in these experiments is formed at the boundary where the local activity is converted into a propagating impulse. This boundary, moreover, is movable. It can be concluded from analysis of the changes in the rhythm of swallowing described above that the disturbance of the normal series of swallowing movements is due primarily to a change in the functional state of the focus of excitation in the swallowing center itself. The gradual increase in latency of the second swallow is probably due to the gradual passage of the swallowing center into a state of inhibition. The presence of inhibition in that particular case is also indicated by the "rebound" observed immediately after disappearance of the second of a series of paired swallows.

An increase in the frequency of swallowing, in the form of paired swallows, by comparison with normal, was observed when the animal's need for water was increased. There is therefore every reason to suppose that this type of phenomenon is principally the result of increased excitability of the swallowing center. The swallowing reflex now takes place in response to a weaker external stimulus, i.e., in response to a smaller volume of water collecting in the animal's mouth in the interval between swallows. After a swallow occurring after the usual interval there is a transient increase in excitability and this leads to the appearance of a supplementary swallow, following a short time after the first.

The fact that the Wenckebach phenomenon arising during rhythmic function of the swallowing center is identical in its principles with the same phenomenon in the human heart and with the phenomenon obtained in experiments on the frog's heart indicates that it is due to changing relationships between the spreading impulse and the focus of stationary excitation. In the case of the swallowing reflex, it must be assumed that the swallowing center itself is this focus of excitation. In other words, the presence of Wenckebach periods is evidence that there must be a rhythmically functioning focus and a block which is in such an unstable functional state that every passing impulse, every response taking place under the influence of this impulse depresses its state toward inhibition, or toward total suppression. In cardiac activity there are individual impulses generated by the cardiac pacemaker; in the swallowing reflex, because of the complexity of innervation of the swallowing apparatus, volleys of afferent impulses travel along the various nerves (trigeminal, superior laryngeal, inferior laryngeal, glossopharyngeal, recurrent laryngeal) innervating the act of deglutition, especially along the first two nerves which are the principal afferent sources concerned in the swallowing reflex (Doty, 1951).

8. Interaction between a Cortical Focus of Excitation and the Swallowing Center

In experiments to study the swallowing dominant the interaction between two rhythmically functioning foci of excitation — the swallowing center and respiratory center — was examined. What type of interaction occurs between two rhythmically functioning foci of excitation if one of them is the swallowing center, but the other is artificially created in the cortex, for example by the local action of strychnine?

As mentioned above, the concept of a dominant focus implies a constellation of effects formed in the course of a response and located at different levels of the central nervous system. It would be wrong to understand by the term "focus" a conglomeration of nerve cells situated, like a nucleus of gray matter, in any one place in the central nervous system. In the swallowing dominant this constellation extends from the medulla to the cerebral cortex.

If a focus of excitation is created experimentally in the motor cortex (the area for the forelimb in the right hemisphere) by application of a piece of filter paper soaked in 0.5-1% strychnine solution, after a short time so-called strychnine waves are obtained in the electroencephalogram (EEG) of this area, indicating the periodic generation of bursts of excitation in the strychninized focus. Judging by the action potentials, periodic spasms develop at the periphery in the corresponding limb.

Does the swallowing dominant inhibit a focus of excitation induced by strychnine in the cortex? Electrophysiological investigations (Fig. 2) show that as soon as swallows began the strychnine

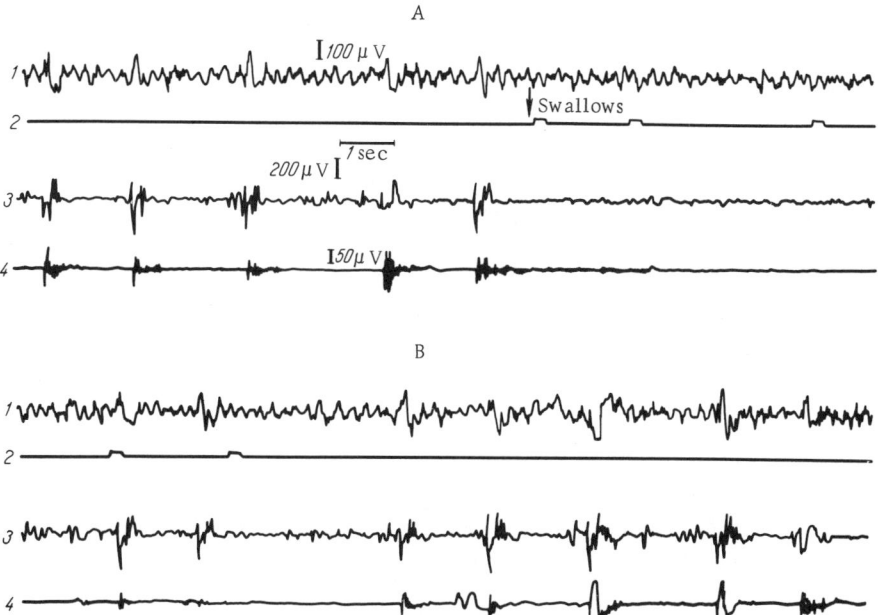

Fig. 2. Excitation of the swallowing center inhibits a focus of excitation in the rabbit cortex induced by 1% strychnine solution (work of Vasil'eva). A) initial record; B) continuation. 1) EEG of area of mastication in the right hemisphere; 2) record of swallowing movements; 3) EEG of right motor cortex (zone for the forelimb); 4) action potentials of muscles of the left forelimb.

spikes in the cortex disappeared. The strychnine continues to act, but the focus of excitation formed by its influence is inhibited. The inhibitory action of the swallowing dominant is particularly conspicuous during the first few swallows. Subsequent swallowing no longer inhibits the focus of excitation. When swallowing comes to an end, the strychnine spikes increase in amplitude compared to those prior to action of the swallowing dominant, and they irradiate more clearly to other regions of the cortex and, in particular, to the cortical area for mastication (Vasil'eva, 1965; Vasil'eva et al., 1966).

These electrophysiological observations show that the cortex participates in the intercentral relationships arising in the presence of the swallowing dominant, that it is a component of the constellation of the swallowing dominant, and that the reciprocal inhibition observed in these experiments takes place at the cortical level.

When strychnine is applied locally to the sensorimotor cortex it creates a focus which, judging from the EEG, passes through several phases. The effect of interaction between a rhythmically functioning focus of excitation (swallowing) and an artificially created focus in the cortex differs depending on the phase of strychnine poisoning in which the swallowing reflex is evoked. The swallowing reflex led to inhibition of the motor effect evoked by strychnine shortly after its application. In the subsequent phases the

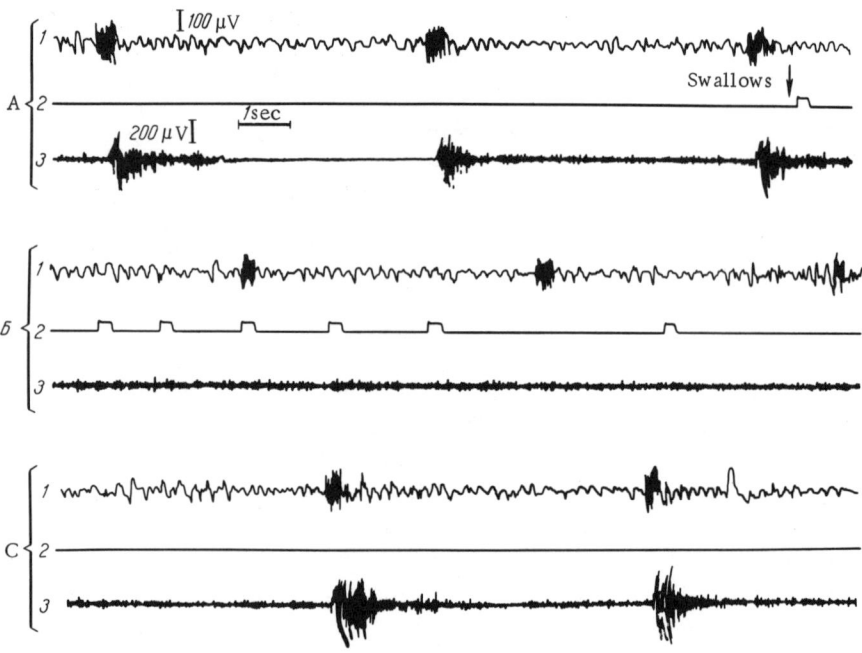

Fig. 3. Swallowing inhibits limb movement from electrical stimulation of cortex (work of Vasil'eva). A) beginning of record; B,C) continuation. 1) EEG of left motor cortex (area for forelimb); 2) swallows 3) EMG from left forelimb. Distance between coils 15 cm. Right motor cortex in area for forelimb stimulated by pulses of induction current.

focus of excitation induced by strychnine begins to acquire dominant properties: the swallowing reflex no longer inhibits, but potentiates the motor effect of the strychnine (Vasil'eva, Nezlina, and Ivannikova, 1966).

The same effect is observed if another method is used. In another series of experiments Vasil'eva stimulated the forelimb area in the right motor cortex locally with weak pulses of induction current and recorded the transcallosal potentials at the symmetrically opposite point of the left hemisphere. In response to periodic stimulation of the right hemisphere contractions of the muscles of the left limb are obtained: periodic action potentials are visible on the electromyogram. The appearance of swallows inhibits the motor effect in the limb despite electrical stimulation of the cortex (Fig. 3).

In these experiments cortical electrical activity differed depending on whether the swallowing dominant inhibited contraction of the forelimb muscles or not. If the forelimb muscles contracted in response to cortical stimulation, the rhythm of the EEG waves was slightly increased after stimulation. If, on the other hand, the response was inhibited, the EEG remained unchanged after stimulation.

How can the gradual weakening of the inhibitory effect of swallows, so clearly visible in Fig. 2, be explained in terms of the dominant? As swallowing takes place, the dominant apparently becomes strengthened, and its effect relative to strychnine stimulation disappears. Gradual weakening of the inhibitory effect in the experiments with application of strychnine to the cortex takes place because the dominant focus itself passes into a state of inhibition during subsequent stimulation.

There is an extensive literature on the action of strychnine on the central nervous system. The view is now widely held that strychnine inhibits the Renshaw cells which, as most workers believe, are inhibitory in function. It is also widely held that strychnine has a specific action by depressing inhibitory synapses. Eccles considered that strychnine depresses the inhibitory mediator at receptor points of the inhibitory subsynaptic membrane and that strychnine under these circumstances exhibits highly specific activity. The view that strychnine acts only on nerve cells and not on nerve fibers was disproved by investigations in Ukhtomskii's laboratory as long ago as the 1930s (Rusinov, 1930a; Konokov, 1930; Golikov, 1933). In particular, it was shown that strychnine, in concentrations of 0.3-0.5%, abolishes the parabiosis produced by the action of KCl in a frog peripheral nerve and restores the function of conduction, but later it actually brings the nerve into parabiosis, which can be abolished, in turn, by Ringer's solution. This sequence can be repeated several times on the same preparation. The effect of local application of strychnine on the central nervous system has been analyzed on several occasions (Vvedenskii, 1906; Beritov, 1917; Beritov and Novinskaya, 1925). The technique of strychnine neuronography, suggested by Dusser de Barenne, is now extensively used to investigate functional links in the central nervous system. Araki (1965) objects to the view, so widely held in the literature, that the mechanism of action of strychnine is to prevent the release of inhibitory mediator in the subsynaptic receptive field. By operating with double microelectrodes in cat motoneurons, injecting ions of different effective diameters, applying a steady current, and using microdoses of strychnine, Araki concluded that strychnine reduces the pore size in the inhibitory subsynaptic membrane. The principal effect of strychnine is depression of the ionic flux through the subsynaptic membrane. In response to a weak concentration of strychnine the pores are reduced in diameter to such an extent that Cl^- ions can pass through the subsynaptic membrane only with difficulty, while K^+ ions can hardly pass through it at all. In response to a high concentration of strychnine, on the other hand, conductance for K^+ ions may be completely blocked, and conductance for Cl^- ions blocked to a considerable extent.

9. Foci of Excitation Induced by Strychnine in Subcortical Structures

I have used strychninization not only to create foci of excitation in various parts of the cortex, but also to obtain foci of excitation in certain subcortical and brain-stem structures, and the results obtained by its use have also served to assess the connection between different structures. Ryabinina (1963), in her experiments on rabbits, showed that if strychnine is injected into the pallidum, high-voltage discharges appear not only in the pallidum

9. FOCI OF EXCITATION INDUCED BY STRYCHNINE

itself, but also in the mesencephalic reticular formation and in the motor cortex. Injection of strychnine into the mesencephalic reticular formation also evoked discharges in the pallidum and in the motor cortex. These observations were made on unanesthetized animals. General anesthesia abolished the paroxysmal discharges.

These investigations point to a functional connection between the pallidum, the mesencephalic reticular formation, and the motor cortex in the rabbit. There is no question about the existence of a two-way connection between the pallidum and the mesencephalic reticular formation, for after strychninization of the pallidum paroxysmal discharges reach the reticular formation and, conversely, injection of strychnine into the reticular formation evokes these discharges in the pallidum. In response to stimulation of the pallidum, just as to stimulation of the mesencephalic reticular formation with strychnine, excitation always spreads to the motor cortex. It can accordingly be concluded that functional connections exist between the pallidum and cortex and also between the mesencephalic reticular formation and the cortex. Both the pallidum and the mesencephalic reticular formation are concerned in the organization of the central motor system.

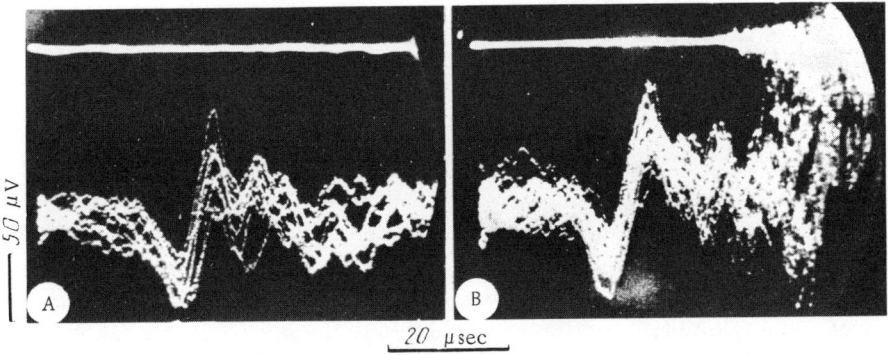

Fig. 4. Appearance of limb movements in response to flashes after injection of 1% strychnine solution into mesencephalic reticular formation of a rabbit (work of Ryabinina): A) before injection of strychnine; B) after injection of strychnine. Top trace shows response of right forelimb; bottom curve evoked response in visual cortex; superposition of 15-20 responses to flash. Pulse generating flash also triggers sweep.

In long-term experiments injection of 1% strychnine into the reticular formation led to the generation of paroxysmal discharges in the reticular formation itself, in the cortex, and in some subcortical structures. At the same time, the animal made generalized movements. Measurement of the latent periods showed that discharges in the motor cortex arise 0.2-0.8 msec later than in the mesencephalic reticular formation. The more caudally the electrode is situated in the mesencephalic reticular formation, the easier it is for paroxysmal discharges to be evoked. During the paroxysmal discharges and for a long time after them the EEG is synchronized, often showing a rhythm of 5 Hz.

The short latency of the cortical paroxysmal discharges suggests that they are presynaptic potentials of the endings of fibers coming from the mesencephalic reticular formation. A role of the cortical pyramidal neurons in the paroxysmal discharges is unlikely in these experiments.

A focus of excitation produced by strychnine in the mesencephalic reticular formation may become a dominant focus. In that case, flashes at frequencies of 1/sec or less evoke a motor response in the rabbit with the same rhythm as the photic stimulus. The latency of the primary evoked response in the visual cortex is 15-20 msec, while that of the motor response is 36-40 msec (Fig. 4).

As the action of strychnine injected into the mesencephalic reticular formation grows weaker, flashes no longer evoke discharges in the cortex, or these discharges arise only in response to the first flashes of a series. However, distinct movements of the animal still continue in response to the remaining flashes, and they still keep strictly to the rhythm of the flashes. After 10-15 min, and in some cases later, the primary responses of the reticular system are diminished, and at the same time the motor responses to the flashes grow weaker. Later only the first flashes of the series still continue to evoke movements. Later still, no movements whatever arise in response either to flashes or to clicks. The focus of excitation induced by strychnine in the mesencephalic reticular formation passes into a state of inhibition under the influence of a series of stimuli. An interruption of a few minutes is sufficient to enable afferent stimuli to evoke "strychnine discharges" and movements of the animal once again.

These responses are then much weaker than at the beginning of action of the strychnine. Return to normal requires 1-2 hr.

These experiments show that strychnine creates a focus of excitation in the reticular formation which, at a certain phase of its development, becomes a dominant focus. It begins to summate excitation from hitherto indifferent stimuli, and it can then close the arc of a motor reflex by means of connections between the mesencephalic reticular formation and extrapyramidal system so that the animal begins to give a motor response to every, hitherto indifferent, stimulus.

These results help to give a physiological explanation of the mechanism of hyperkinesia, a phenomenon which closely resembles a pathological dominant, and which bears further examination. The globus pallidus, which receives afferent impulses from the reticular system of the brain stem can, in turn, have a reciprocal action on this system. Besides corticostriatal connections, there are also pallido-cortical connections, which are responsible for one of the feedback mechanisms in cortico-subcortical relationships. This "loop" can be the source of formation of a focus of stationary excitation via cortical—pallidal—cortical connections. This central focus of excitation, connected as it is with the "loop" of reticular system of the brain stem—pallidum—reticular system, or with the longer "loop" described in the literature, taking in the diffuse system of the thalamus and other basal ganglia, can direct the response of the organism toward a particular reflex, regardless of the place or character of the stimulus, for a considerable period of time and, in hyperkinesias, virtually at all times. Under pathological conditions, with inertia of stationary excitation in the focus, through its connections with the reticular system it can give rise to tremor and muscle rigidity, i.e., it lies at the basis of the "mechanism" of hyperkinesias.

The physiologists O. M. Grindel' and S. N. Raeva, in collaboration with the neurosurgeon E. I. Kandel' (Grindel' et al., 1962), during stereotaxic operations on the basal ganglia investigated changes in the EEG of patients with parkinsonism. The operations were intended to destroy the medial part of the globus pallidus or the ventrolateral nucleus of the thalamus. The EEGs were analyzed by means of a frequency analyzer and electronic integrator.

The investigation showed that in most patients with parkinsonism the EEG remains the same as in the preoperative period, despite the fact that the operation gave clinically good results. This shows that the operation removed the principal manifestations of the disease (the tremor and muscle rigidity) evidently by breaking the pallidum–diffuse thalamic system–mesencephalic reticular system–pallidum loop, but did not eradicate the diffuse pathological process in the basal ganglia (perhaps in the substantia nigra), which gives rise to the pathological changes in the cortical EEG secondarily.

10. A Focus of Excitation in the Visual Cortex

What is the relationship between the focus of excitation in the cortex and the inflow of afferent impulses? A formed focus cannot be a simple effector agent, responding to the arrival of impulses and whose response depends automatically on the quantity and quality of the incoming information. It actively regulates the flow of afferent impulses through a system of feedbacks, of corticofugal influences. These influences of the cortical focus of excitation evidently are felt by the afferent impulses along the whole length of their path, but in particular in their relay stations.

In our laboratory Novikova and Farber (1956) showed that frequent photic stimuli applied to one eye of a rabbit lead to the formation of a focus of excitation, with the properties of a dominant, in the contralateral occipital region. A focus of excitation with dominant properties can also be created there by polarization with a weak, direct current. The most interesting fact revealed by their experiments was that changes in electrical activity of the retina take place in response to acoustic reinforcement of the dominant focus situated in the occipital region of one hemisphere. The visual system behaved in these experiments as a single functional formation in which the functional state of the retina is under corticofugal control. With intensification of the focus of excitation as a result of prolonged action of the direct current or an increase in its strength, periods of spreading depression from the region of polarization to other regions of the same hemisphere are observed, and, it must be emphasized, during the depression of electrical activity of the focus of excitation in the occipital region, electrical activity of the retina is also depressed. It thus follows that if a focus of excitation is present in the visual cortex, not only excitatory, but also inhibitory influences spread from the cortex to the retina along retrograde connections. After division of the optic nerve, the rhythmic retinal potentials in experiments to create a focus of excitation in the visual cortex are no longer recorded in response to acoustic stimulation. These experiments reveal the physiological mechanisms of the rhythmic waves recorded in the retina during acoustic stimulation. After division of the optic nerve, the influences of the focus of excitation as feedback from the cortex to the retina disappear.

11. Corticofugal Influences from Foci of Excitation

What are the connections which transmit such corticofugal influences?

The corticofugal fibers in the visual system run to the superior colliculi, to the pulvinar, to the pretectal zone, and to the lateral geniculate body (Elinson, 1896; Dejerine, 1901; Kononova, 1926; Shkol'nik-Yarros, 1958).

Dzugaeva (1958) found direct fibers connecting the retina and cerebral cortex in man. Direct connections between the peripheral component of the visual system and the cortex, in her opinion, are provided by both centripetal and centrifugal fibers.

The problem of corticofugal connections with the various subcortical structures has recently been extensively studied in both Soviet and Western laboratories. However, the results described are contradictory. Some aspects of this problem have been dealt with by Narikashvili (1962), Gustson (1964), and Meshcherskii (1964, 1966).

This problem is relevant here in connection with cortical foci of excitation. Cortical modulation of the conduction of afferent impulses through the thalamic relay nuclei may be one of the mechanisms whereby a focus of excitation is converted into a dominant.

Meshcherskii and Gustson (1964), in our laboratory, showed that during the unilateral spread of depression over the rabbit's cortex the amplitude of the primary responses to flashes is reduced in the ipsilateral lateral geniculate body. These findings show that the cortex exerts local control, via corticofugal pathways, over the thalamic relay via recurrent corticothalamic connections.

Influences from the projection zone of a particular hemisphere are exerted only on the ipsilateral lateral geniculate body, i.e., they are strictly localized. Intravenous injection of 0.2 ml 0.1% strychnine solution does not prevent the decrease in amplitude of the responses of the lateral geniculate body during a wave of spreading depression. The same results were obtained when strychnine was applied to the cortex. It was next shown in experiments on cats that after coagulation of the visual area II (V-II) and application of penicillin to visual area I (V-I), transcallosal stimulation of V-I evokes responses both in the lateral geniculate body and in the middle part of the marginal gyrus. The same stimulation after coagulation of the primary zone does not evoke responses (Meshcherskii, 1966).

These results indicate that besides influences from the reticular system, which are general, and familiar from the literature, there are other mechanisms giving rise to a more local effect. In particular, these influences in the visual system are exerted on the thalamic relays directly via the centrifugal specific pathway, and it is very possible that they are one of the mechanisms which maintain, through feedback, the level of excitation in the focus as the optimal level for the particular state of the system. Whether this feedback is inhibitory or facilitatory depends on a number of factors and, in particular, on the level of excitation in the focus itself.

* * *

In sum, to the four known features of the dominant focus must be added another — disinhibition of the focus. The dominant is merely one stage in the functional evolution of the focus of excitation in the central nervous system. It is one form of activity of foci of excitation in which the focus is established at a certain level of stationary excitation, permitting the summation of excitation reaching the nervous system and the creation of optimal conditions for operation of the focus, so that it is primarily this focus which responds to afferent stimulation and other active foci are correspondingly inhibited.

During rhythmic activity of the swallowing center Wenckebach's phenomenon arises, and is identical in its underlying principles with the same phenomenon in the human heart and the phenomenon obtained in experiments on the animal heart. It is due to changing relationships between the propagating impulse and the focus of stationary excitation.

Cortical modulation of the conduction of afferent impulses through the thalamic relay nuclei may be one of the mechanisms whereby the focus of excitation is converted into a dominant. Besides generalized influences from the reticular system, there are also mechanisms exerting a more local effect. In particular, these influences in the visual system are exerted on the thalamic relay directly via specific centrifugal pathways, and they probably act as a feedback mechanism maintaining the level of excitation in the focus at the optimal level. Whether the feedback is inhibitory or facilitatory is largely determined by the level of excitation in the focus itself.

Chapter II

The Problem of Stationary Excitation and Changes in the Cortical Steady Potential during the Formation of Dominant Foci and Temporary Connections

1. Two Approaches to the Theory of Excitation

Anyone who examines the history of development of theories of excitation in the general physiology of the nervous system will detect two clearly distinguishable approaches. One is well covered in the world physiological literature. This is the approach developed some years ago by the Cambridge school of physiology on the basis of the results of experiments on peripheral nerve. The followers of this approach regard every type of excitation as a propagating impulse reflected electrographically as the action current of a traveling wave. The older generation of English physiologists regarded this impulse as the standard or unit of excitation. This view was put forward by Adrian, who applied it to the physiology of the central nervous system, as is clear from his explanations of the slow waves in the electrocorticogram. Adrian (Adrian and Matthews, 1934) considered that all waves of electrical potential in the cortex are formed from relatively short pulsations in single neurons and that slow changes of potential result from summation. This means, Adrian wrote, that hope must be abandoned for the idea that slow changes of potential in the cortex can indicate a slow change of polarization in nerve cells or indicate the rise and fall of excitation which is reflected in this activity. Many slow changes of potential observed in other structures of the central nervous system can be due, in his opinion, to slow changes in the cortex, to the summation of fast action currents repeating at a frequency which varies within wide limits. He had already begun to explain his earlier findings in connection with slow changes of potential and, in particular, with the "respiratory waves" in the goldfish brain stem (Adrian and Buytendijk, 1931) and in the thoracic and optic ganglia of Dytiscus (Adrian, 1931, 1932), not as previously by gradually developing depolarization, but just as in the cortex, by summation of a series of short impulses which may vary in frequency, but not in amplitude or shape.

The other approach, which has received less attention in the world physiological literature, was conceived originally by Vvedenskii (1901) and later by Ukhtomskii (1932) in a theory of stationary excitation also based on results obtained primarily from investigations on peripheral nerves, but allowing a different interpretation. These workers considered that excitation, although remaining a single process in every case, can manifest itself in various modifications depending on the structure undergoing excitation and on its functional state. The moving impulse, reflected electrographically as an action potential, spreading in a wave along nerve fibers, is not the only manifestation of activity of the nervous system. There is another form of nervous activity, namely stationary excitation, which can be reflected electrographically only as slow, long-lasting potentials.

2. Slow Long-Lasting Potentials in the Cortex

Slow, long-lasting potentials have now been demonstrated in the cortex by several workers. The amplitude and direction of these potentials depend on the different methods used, but the fact that such long, slow potentials exist in the cortex is no longer in doubt (Rusinov, 1947, 1949, 1962; Lur'e, 1949; Lur'e and Rusinov, 1955; Beritov and Roitbak, 1955, 1957; Shvets, 1958, 1960a, 1960b, 1963; Aladzhalova, 1960, 1962; Mnukhina, 1961; Shuranova, 1966; Morrell, 1961, 1962a, b; Goldring and O'Leary, 1951, 1954, 1957, 1960; Arduini, Mancia, and Mechelse, 1957; Bureš, 1957; Brookhart, Arduini, Mancia, and Moruzzi, 1958, and others).

Long-lasting electrical potentials in the cortex are described in the literature under many different names: slow, long-lasting potentials, aperiodic slow waves, changes in level of the steady potential, ultraslow waves, smooth potentials, and so on; they define bioelectrical processes which are usually longer than the delta-waves of the EEG (over 1 sec). Although phenomena of this character have long been known, even longer than the action potential complex and the EEG with its rhythmic waves, they have only achieved recognition in the world literature comparatively recently, during the last 10-15 years.

According to Arduini et al. (1957) and Goldring (1958) afferent impulses from the thalamus and ascending reticular formation of the brain stem displace the steady cortical potential toward negativity. According to Brookhart et al. (1958), the negativity associated with stimulation of the thalamic specific nuclei is limited to a definite region of the homolateral cortex, whereas during stimulation of the intralaminar thalamic nuclei negativity is observed in both hemispheres, and it reaches a maximum in the fronto-central regions. Stimulation of the mesencephalic reticular formation differs from stimulation of the medial thalamus, for it spreads to a wider area including the parietal and occipital regions. The amplitude and shape of practically all potentials recorded from the brain surface depend on the height and level of the cortical steady potential.

Arduini, Mancia, and Mechelse (1957) showed that enduring changes of potential are observed widely over the cortex in response to high-frequency stimulation of the mesencephalic reticular formation. Negative deviation of the slow potential (200-400 μV) continues for 1-2 sec after the end of stimulation and is recorded over the whole of the dorsal surface of both hemispheres, about equally at symmetrical points. This potential is neuronal in origin, for it increases considerably after local application of 0.1% strychnine solution and disappears completely after local application of 0.6% nembutal solution. Recordings of the EEG with a dc amplifier in preparations (such as the "encéphale isolé" cat) with a well defined basal rhythm of electrical activity showed that the threshold parameters for evoking desynchronization and a slow long potential are equal, and that both these phenomena appear at the beginning of stimulation. These workers also found that repeated stimulation of afferent nerves causes the appearance of a slow potential over a wide area of the cortex. They consider that slow changes of potential recorded from the cortical surface depend on excitation of the reticular formation.

The background gradient of the cortical steady potential is also a well established fact. Longitudinal, transverse, and transcortical gradients of steady potential have been found in several types of preparation (Libet and Kahn, 1947; Goldring and O'Leary, 1951; Gumnit, 1961). Opinions differ regarding the origin of the gradients of steady potential level (SPL) in the cortex. The view of Gerard and Libet that these gradients arise through inequality of polarization of different parts of the neuron, is widely held. Investigations on single-layered hippocampal structures have shown that the bodies of pyramidal cells are actually negative relative to their apical dendrites. However, in embryos, in which the subdivision into cells and dendrites is not yet clearly defined, the cortex is positive relative to an indifferent point (Eidelberg et al., 1965).

Administration of several pharmacological agents, a change in the blood pH, and sleep, all significantly change the initial SPL (Caspers, 1964; Kawamura and Sawyer, 1964; Wurtz, 1965; Tabushi, 1965).

Changes in SPL can be induced by direct action on the cortex — by electrical stimulation, by pressure (Chang, 1951; Keating and Kempinsky, 1960; Shvets, 1960; Mingrino et al., 1963), and by application of various substances (Goldring and O'Leary, 1954).

These results show that steady potentials in the central nervous system are a bioelectrical phenomenon which is just as widespread as electrical fluctuations of other parameters, or perhaps even more so.

The results obtained using sensory stimuli of different modalities or electrical stimulation of afferent fibers are contradictory both as regards the degree of localization of recorded changes in SPL in the cortex and also as regards their sign. Some consider that stimulation of various modalities gives rise to local responses (amplitude 50-1000 μV, duration several seconds) in the corresponding projection zones (Köhler and O'Connell, 1957; Goldring and

O'Leary, 1960). Others conclude that the cortical response is diffuse, and is most marked in the frontal regions (Cowan et al., 1965).

A local change in SPL has been observed principally in investigations on cats and dogs, and a diffuse change in investigations on rabbits. However, there is some evidence that in cats and in man the change in SPL in response to photic stimulation may be more marked (Cowan et al., 1965; Walter, 1964, 1965) in the frontal regions. Shvets, investigating steady cortical potentials in our laboratory, found that the results depend on the experimental conditions. In experiments with platinum electrodes implanted on the dura in rabbits, responses were obtained only from electrodes from which a high (ranging from several millivolts to several tens of millivolts) negative initial SPL had been recorded. These responses were independent of stimulus modality and electrode location. They consisted of a negative wave, very variable in amplitude (from 500 μV to several millivolts) and duration (from 1 sec to tens of seconds). The responses appeared very regularly, being recorded to 50-90% of applications of the stimuli. However, virtually no changes in SPL (only 3-5%) were recorded in the same animal under the remaining electrodes (up to 10 of them) in response to application of various stimuli. A different picture was observed in experiments in which the changes in SPL were recorded by calomel or $Zn-ZnSO_4$ electrodes from the thinned calvarium.

In these experiments photic stimulation evoked changes in SPL from 30-80% depending on the strength of stimulation. The changes were diffuse, were recorded from the occipital, parietal, and frontal regions, and were both positive and negative. The amplitude of the changes reached 300-500 μV.

In rabbits, but under different experimental conditions, different results were thus obtained, for response changes in SPL from electrodes implanted on the dura were local and negative in sign. However, the degree of localization was determined not by the modality of the stimulus, but by the functional state of the tissue under the electrode. In experiments in which the electrodes did not touch the brain the response changes in SPL were diffuse and were either positive or negative in sign. It was postulated that in the first case a focus of excitation with a high initial negative SPL was formed, possibly as a result of pressure from the electrodes, in the surface layers of the cortex under the electrode. Shvets then deliberately created a focus of pressure in this way and obtained positive results: a high negative SPL in the region of pressure and an increase in this negative SPL in response to various forms of stimulation. Virtually no changes in SPL were obtained under any of the other electrodes. These results suggest that changes in SPL are very sensitive to the experimental conditions, more sensitive than various types of evoked potentials or the EEG. This has also been observed by others (Brookhart et al., 1958; Caspers, 1961; Gumnit, 1961) in connection with levels of narcosis at which changes in SPL, changes in body temperature, and changes in the general functional state of animal can be observed. It is probable that more complex neural mechanisms underlie the changes in SPL.

In experiments with the intact cortex, when the SPL was recorded by nonpolarizing electrodes through a thinned area of the calvarium (to the lamina vitrea, diameter 0.5-0.7 mm), Shvets showed that the regularity with which changes in slow potential appear as responses, their predominant sign, and their distribution over the cortex depend on the strength of stimulus applied. During weak stimulation changes in SPL with a positive sign were predominant. With an increase in the strength of stimulation, negative changes became sharply predominant.

If the view is held that changes in SPL in nonprojection zones with respect to stimulation of a certain modality are evoked by impulses traveling along nonspecific pathways, while those in projection zones are due to impulses traveling mainly along specific pathways, it follows from these experiments that both specific and nonspecific changes in SPL can be both positive and negative in sign. So far as the nonspecific changes are concerned this conclusion does not agree with the views of those who consider that nonspecific afferent fibers terminate mainly on apical dendrites, and that excitation arriving by these fibers evokes their depolarization, i.e., a surface negative change in SPL. However, results have been described which indicate that during electrical stimulation of nerve trunks or of various nonspecific nuclei the polarity of the change in SPL may depend on the strength, duration, and frequency of stimulation. By changing the parameters of stimulation both negative and positive deflections of SPL can be obtained (Goldring et al., 1958; Brookhart et al., 1958; Bonnet, 1958).

To judge from the results obtained by Shvets with the use of sensory stimulation, surface-positive zones predominate at low strengths of stimulation for both nonspecific and specific changes, whereas surface-negative zones predominate for strong stimulation. In an attempt to explain this phenomenon Goldring and O'Leary (1951, 1957) postulated that changes in SPL of the cortical surface are complex in origin and are the result of algebraic summation of a series of fields of steady potential with opposite signs. With a change in the parameters of the stimulus or in the site of stimulation or recording, the amplitude of the various fields and, correspondingly, the sign of the recorded SPL field, also change. What is the source of these fields — depolarization or hyperpolarization? In what layers of the cortex does this take place? These questions still remain unanswered.

There is another interesting fact: with an increase in stimulus strength the "center of gravity" of the changes in SPL shifts from the posterior to the anterior regions of the cortex. The results of investigations into changes in cortical surface SPL during stimulation of various thalamic and mesencephalic nuclei show that in response to stimulation of the nonspecific nuclei the most marked changes in SPL are found in the anterior regions of the cortex, while in response to stimulation of specific nuclei they are found in the corresponding projection zones and near to them (Arduini et al., 1957; Goldring and O'Leary, 1957; Bonnet, 1958; Brookhart et al., 1958; Vanasupa et al., 1959). Hence, in response to a flow of afferent impulses, fields of steady potential, differing in both origin and direction, arise both in the projection zone and in other regions of the cortex. In response to relatively weak stimuli positive fields predominate and most of the changes take place in specific regions; with an increase in the strength of stimulation negative changes become predominant and most of the changes take place in the anterior region.

3. Interaction of Slow Potentials with Evoked Responses and Dendritic Potentials

If the appearance of a slow potential in the cortex depresses the basic EEG rhythm, the question arises: how will it affect responses evoked by afferent stimuli?

Stimulation of the reticular formation affects not only spontaneous cortical activity, but also evoked potentials. Desmedt and La Grutta (1957) showed that stimulation of the reticular formation usually reduces the amplitude of primary responses to single stimuli applied to the medial geniculate body. Others have shown that primary responses are blocked, while secondary responses are sometimes reduced, and the aftereffect following the secondary discharge is blocked constantly.

All these evoked cortical potentials (primary and secondary responses, "recruiting" or "augmenting" responses) appear in the cortex to stimulation of extracortical structures. Evoked potentials of a different type, known as surface (Adrian, 1941) or dendritic (Chang, 1951; Bishop and Clare, 1953) responses to stimulation of the cortex itself at a distance of 2 mm from the recording electrodes are also blocked during stimulation of the reticular formation (Purpura et al., 1956).

Short electrical stimulation of low intensity evokes a slow potential on the cortical surface, the chief components of which are two surface-negative waves differing in their latent period, amplitude, and duration. In response to stronger stimulation, a surface-positive potential of much lower amplitude may appear before the negative waves. Usually these potentials are described as dendritic potentials. Whether these potentials develop entirely in the apical dendrites, as some suggest, has not yet been precisely demonstrated. Very possibly they reflect changes in potential of the presynaptic fibers and axo-dendritic synapses also. This problem has been examined in greater detail by Okudzhava (1963).

Purpura and Grundfest (1956) maintain that dendritic potentials are postsynaptic potentials. They concluded from pharmacological experiments that structurally fixed types of synapses are responsible for the functions of excitation and inhibition. Others point to a dynamic relationship between responses to cortical stimulation and various factors, notably the long-lasting, slow potential on the cortical surface.

The relationship between slow, long-lasting potentials and dendritic potentials has been investigated by Caspers and Schulze (1959). They showed that the amplitude, shape, and polarity of the dendritic potential consistently depend on the level of the steady potential of the cortical surface. If the cortex is polarized by an anode or by local application of substances producing positivity, the surface-negative dendritic potentials become much greater in amplitude and duration. The change toward negativity of the cortical steady potential during cathodal polarization, during a series of weak direct stimuli, or local application of γ-aminobutyric acid acts in the opposite way — it lowers the level of the steady potential and terminates with disappearance of the dendritic response. The level at which the dendritic response disappears completely is defined as the "indifferent level" of the steady potential. An increase in negativity of the cortex causes the dendritic potential to disappear on reaching the indifferent level, after which, changing in direction, the potential again increases its amplitude.

Investigations into these changes in the cortical steady potential during natural periods of sleep and waking have shown that the cortical steady potential depends on the level of the animal's activity. During sleep the steady potential shifts toward positivity. Awakening produces negativity (Arshavskii, 1956; Caspers et al., 1959), and this increases in intensity if the animal moves. Shvets (1963a), found that an increase in, or appearance of cortical negativity during movement is observed not only during sleep—waking periods, but also in the waking state; before a rabbit's spontaneous movement, the steady potential changes toward negativity.

Slow potentials have recently attracted increasing attention on the part of electrophysiologists. Evidence is steadily accumulating that slow potentials, differing in their characteristics from axon potentials, reflect important neuronal functions. Investigations showing the appearance of a slow response in the cortex to afferent stimulation and to stimulation of the reticular system have led, in particular, to further attempts to elucidate thalamo-cortical relationships by studying changes in slow potentials in the cortex in response to stimulation of various thalamic nuclei. Stimulation of some thalamic nuclei by a low-frequency (7-12/sec) electric current evokes responses at the cortex which increase progressively during the first 3-6 stimuli.

Dempsey and Morison (1942), who discovered this phenomenon, observed two types of response, differing from each other in their distribution over the cortex: responses to stimulation of specific thalamic projection nuclei (augmenting responses) and responses to stimulation of the diffuse thalamic system (recruiting responses).

Brookhart, Arduini, Mancia, and Moruzzi (1958) investigated slow potentials in the cat cortex for stimulation of nuclei of the specific and diffuse systems of the thalamus, evoking the phenomena of Dempsey and Morison. With stimuli at 6-10/sec the "augmenting" or "recruiting" responses disappeared if the stimulating electrodes were moved vertically through a distance of 1-2 mm, as had previously been shown by the experiments of Jasper (1960) and Pavlygina (1963). In recordings made with a dc amplifier both types of responses began on the base line, and shifted gradually toward negativity during the course of stimulation. If the frequency of stimulation reached 40-60/sec, the negative potential arising on the brain surface was more prolonged, lasting 1.5-2 sec after the end of stimulation. During electrical stimulation, whether of low or high frequency, the change in the slow potential was always negative relative to the distant electrode.

Comparison of responses on the cortical surface to stimulation of the nonspecific and specific thalamic nuclei shows that different responses are obtained in the anterior zones. In response to stimulation of medial zones of the thalamus, bilateral slow potentials are recorded. If the stimulation is at low frequency (8/sec), a recruiting effect is obtained against the background of the slow potential, and it also is bilateral. In response to stimulation of the lateral thalamic nuclei the slow negative potential and the "augmenting" response appear in the cortex only on the side of stimulation. Slow potentials in the anterior cortical regions are bilateral: they appear in response to stimulation of the medial thalamus and also to stimulation of the reticular formation. In the posterior zones of the cortex, on the other hand, a slow potential appears only in response to stimulation of the mesencephalic reticular formation.

The optimum frequency for thalamic stimulation lies within the limits of 50-70/sec, whereas the optimum frequency for stimulation of the reticular formation for evoking a slow potential is about 100/sec.

Where does this slow long negative potential arise: in the cortex or somewhere in the subcortical structures?

The intracortical, neuronal origin of the slow potentials was demonstrated by Brookhart et al. by local chemical and electrical stimulation changing the state of the cortex. The slow negative potential was reduced, and sometimes changed toward positivity in response to surface-negative polarization (0.3 mA) and was increased by surface-positive polarization. These effects were completely reversible. When the polarizing electrode was moved a few millimeters away from the recording electrode, no such changes took place in the magnitude of the slow potential.

Experiments in which a 0.1% strychnine solution was applied locally to the recording point give further evidence of the cortical origin of the slow potential. Local application of strychnine of the same concentration a few millimeters away from the recording point, however, does not increase the voltage of the slow potential.

After extirpation of the cortex at the point where a slow potential is recorded and replacement of the defect by moist cotton no slow potential appears in response to subcortical stimulation.

These results are evidence of the cortical localization of the physiological mechanisms generating the slow potential and also of the cortical localization of the functions reflected electrographically as slow potentials. The Italian investigators consider them to depend on the postsynaptic activity of cortical neurons evoked by the arrival of thalamo-cortical impulses. However, the facts described above are indirect evidence rather of the neuronal nature of the changes. Experiments in which the changes in SPL and changes in membrane potentials were correlated would provide more direct and convincing evidence.

4. Hypotheses on the Genesis of Slow Electrical Potentials

The hypothesis on the genesis of slow cortical potentials which merits the closest attention is that of Libet and Gerard (1941), according to which changes in SPL arise from the degree of polarization of the cortical cell membranes. The view has been expressed that slow electrical potentials are produced by summation of postsynaptic potentials. It has been postulated that synaptic potentials themselves are heterogeneous: some potentials are faster and some are slower (Goldring et al., 1959, 1960; Pearlman et al., 1960;

Roitbak, 1965). Summation of the slower synaptic potentials gives rise to the steady potential. The possibility likewise cannot be ruled out that slow cortical potentials are analogous to electrotonic spinal potentials, i.e., that they reflect prolonged presynaptic depolarization (Eccles et al., 1962).

The suggestion has also been made that the steady potential is generated in the cortex by the membrane potential, not of the neurons, but of the glia. Evidence in favor of this hypothesis is given by experiments with glial cell cultures in which, during electrical stimulation of the glia, long changes in membrane potential were recorded (Hild, Chang, and Tasaki, 1958).

Galambos, after reviewing the published data on the action of steady current on the cortex and, in particular, on the ability of a weak steady current acting over a long period to alter the functional state of neuronal activity (Rusinov, 1953), considers in the light of his hypothesis of the "glial-neuronal function of the brain" (1961) that these facts, which are directly related to the formation of temporary connections, cannot be explained entirely by changes taking place on the neuron membrane. Galambos regards the glia and neurons as functional units working in concert and he postulates, as do others (Kuffler and Nichols, 1966a,b; Roitbak, 1965, 1970) that the glial cells are possibly responsible for slow waves. Although this hypothesis put forward by Galambos by no means rests on a factual basis, it can be useful for further research.

Tasaki, Chang, and others have shown that astrocytes in tissue culture respond to stimulation by generating electrical activity, and this activity is slow in nature. Impaled astrocytes in tissue culture in the experiments of Hild, Chang, and Tasaki (1958) gave an electrical response to direct stimulation through an extracellular electrode which was more than 100 times longer in its duration than the action current of the nerve cells in the same culture. Tasaki and Chang (1958) described results indicating that the glial cells in the mammalian cortex can also give an electrical response to direct stimulation. Ultramicroelectrodes in the cat cortex recorded slow, reversible changes of potential reminiscent of electrical responses from the glial cells in tissue culture.

These results have shed light on some new problems in the physiology of the central nervous system and, in particular, on the investigation of the physiological mechanisms of temporary connections.

To determine the properties of glial cells in the vertebrate central nervous system a suitable preparation was needed. Kuffler et al. (1966) showed that the amphibian optic nerve contains relatively large glial cells, which directly surround groups of nonmedullated axons. Impulses passing along the nerve fibers induce depolarization in the surrounding glial cells.

All that is clear at present is that the SPL in the cortex and its changes constitute a complex phenomenon which is the result of summation of many electrical fields of different origin and direction.

What is debatable in the work of the Italian writers cited above is their explanation of the mechanism of origin of the slow negative potential. They state that when they applied stimuli with gradually increasing frequency, the individual waves in response to each stimulus progressively diminished, being apparently "submerged" in the increasing negativity of the slow potential. The resulting electrographic records show, they consider, that the "building blocks" of the smooth slow potential during stimulation (6/sec) are the surface-negative components of each evoked response. In other words, the slow potential is the sum of these components. The temporal course of the so-called dendritic potentials is such that they can be summated at frequencies used in the experiment. This explanation is contradicted by results obtained by others showing that slow negative potentials on the cortical surface are obtained without interrupting stimulation of any subcortical structures whatsoever, when no evoked responses are present.

Almost simultaneously with the paper by Brookhart et al., or in fact a little before it, Goldring and O'Leary (1957) published their investigation in which they found changes in the slow potential, or steady potential, in the cortex in response to stimulation of the midline of the thalamus. Starting in 1951, Goldring and O'Leary have systematically studied problems connected with the cortical slow potential. Their earlier investigations (Goldring and O'Leary, 1951, 1954) revealed aftereffects from an evoked response and from strychnine and veratrine spikes. These aftereffects can be summated, as they observed after the application of various stimuli to the optic nerve and lateral geniculate body. The polarity of this afterpotential is the same as that of the principal phase of the evoked response or of strychnine or veratrine spikes. A negative potential thus follows the negative spike resulting from the action of strychnine, and a positive potential follows the veratrine spike. In the case of a biphasic evoked response, the aftereffect may also be biphasic.

The relationship between the afterpotential and evoked responses in the cortex has not yet received an adequate explanation. The slow potential of the aftereffect may be an active neuronal process which differs from that reflected in evoked responses just, for example, as does the aftereffect in peripheral nerve; it may be the effect of slow recovery by repolarization, or there may be other explanations.

Goldring and O'Leary conducted their investigations on rabbits and cats. They showed that the cortical slow potential is obtained in response to midline thalamic stimulation both in rabbits and in cats, although there are certain differences. In response to stimulation at a frequency evoking "recruiting" in the rabbit, a negative potential appears. In the cat there is either no slow potential whatever in response to stimulation at that frequency, or a small positive potential. In response to high-frequency midline stimulation of the thalamus in the rabbit a slow negative potential with an amplitude of up to 1.5 mV is obtained, which continues for a few seconds after the end of stimulation. In the cat a negative slow potential is obtained in response to high-frequency stimulation, but it is less marked than in the rabbit.

Goldring and O'Leary consider the slow potential appearing in the cortex in response to low-frequency midline thalamic stimulation to result from summation of aftereffects remaining in the cortex from each wave arising in response to stimulation. During high-frequency stimulation the slow potential, in their opinion, is the result of summation not only of aftereffects, but also of evoked responses themselves. The whole result of summation is regarded as a steady state of dendritic activity. This view is based on the result of an investigation of dendritic potentials by Clare and Bishop (1955), who described the dendritic origin of recruiting potentials. When they studied the properties of apical dendrites they found deviation of the slow potential resulting from the summation of dendritic potentials in response to rapid stimulation. When the second stimulus follows the first after an interval of 20 msec the response to the second stimulus may arise at the beginning of the descending phase from the preceding stimulus. In other words, the dendrite has no absolute refractory phase, and as a result summation takes place. In response to high-frequency stimulation a smooth curve is recorded as a slow potential.

Goldring and O'Leary obtained results which show that a positive slow potential usually appears at the cortex in response to stimulation of nuclei of the specific system (at high frequency). These results do not agree with those obtained by Brookhart and by the Italian workers.

5. Infraslow Rhythmic Potentials

Unlike the authors cited above, Aladzhalova (1962) investigated very slow rhythmic potentials with a frequency, in most cases, of between 0.5 and 8/min. To be more precise, by the term "infraslow rhythmic potentials" she referred to nearly sinusoidal potentials, with a frequency of 7-8/min and amplitude 0.3-0.8 mV (second rhythm) and with a frequency of 0.5-2/min and amplitude 0.5-1.5 mV (minute rhythm). Aladzhalova's investigations showed that intermediate frequencies also exist. The very slow potentials found by Aladzhalova in the brain of warm-blooded animals are rhythmic, i.e., they differ from the slow deflections of cortical electrical potential which have been analyzed so far in this book and which, in most cases, are aperiodic.

Aladzhalova's investigations showed that the infraslow rhythmic waves are manifested differently in different brain structures. They are observed mainly in the cortex and hypothalamus, and are absent in the thalamus of intact animals and in the reticular formation of the brain stem. They are not observed everywhere in the cortex, but occur mainly in the region of the apical dendrites and bodies of the neurons in layers V—VI.

Investigation of the basic principles governing infraslow rhythmic potentials showed that different frequencies can be recorded in different zones of the cortex. Definite differences were found between the sensorimotor and visual areas. Recordings made from different levels of the cortex also revealed rhythms of different frequencies. Aladzhalova accordingly concluded that the processes reflected in infraslow rhythms are local. In experiments in which the blood pressure was varied she showed that there is no direct mechanical connection between the cerebral hemodynamics and these very slow rhythms. After brain trauma, or after electrical and mechanical stimulation of the cortex, the infraslow rhythms disappear. After sympathectomy, changes in the electrical conduction properties of the cortex are accompanied by an increase in the intensity of these waves. In the initial, superficial stage of anesthesia, infraslow waves in the upper layers of the cortex are regular in form. During deep anesthesia these waves disappear completely. Two types of connection were found between the infraslow and faster waves. The first type is based on periodic changes in excitability of the cortical cells synchronized with the infraslow waves, while the second is based on the conditions determining synchronized activity of the cortical neurons. This last connection operates in the case of stimuli evoking a generalized effect in both hemispheres and involving lower levels of the brain in the response. Facts relating to the physiological mechanisms of these types of connection were obtained by studies of evoked potentials and recruiting responses and by direct stimulation of the cortical surface.

In some interesting experiments Aladzhalova isolated a strip of cortex so that it had no neural connections with neighboring areas or with the white matter, but it retained its blood supply through the pia. This preparation, originally described by the British physiologist Burns, was used by Aladzhalova to distinguish between the aperiodic waves of steady potential and the infraslow rhythms which she had found. In the resting state of the strip the faster electrical activity disappears whereas the infraslow waves remain. She accordingly concluded that infraslow waves are observed independently of the presence or absence of dendritic potentials and of other forms of faster electrical activity. In her opinion the infraslow waves themselves reflect processes determining periodic changes in excitability of the neurons. The nonspecific brain system is recruited into the physiological mechanism of augmentation of the infraslow rhythms. In particular, stimulation of the dorsomedial nucleus of the hypothalamus affects the amplitude and regularity of the infraslow waves, whereas stimuulation of the reticular formation of the brain stem and thalamic nuclei does not give this effect. Only following alteration of the metabolism of these structures can infraslow waves be found in the reticular formation of the brain stem and in the thalamus. Aladzhalova's findings indicate that the hypothalamus plays a particularly important role in the genesis of infraslow potentials.

6. The Problem of Stationary Excitation

Ukhtomskii regarded the investigation of stationary excitation as one of the principal tasks facing physiologists studying the central nervous system. His views are given in his address to the Academy of Sciences of the USSR (1932): "The problem of stationary excitation."

Vvedenskii and Ukhtomskii reached a conclusion of great importance at that time, and one which enabled them to erase the traditional sharp line between the concepts of resting potential and action potential. They asserted that both potentials are associated with an active physiological state. The functional role of stationary local excitation, in the form of foci of excitation, would be absolutely confirmed by the existence of slow potentials, whose discovery in the central nervous system was predicted by these physiologists. They set out from their theory, which was aimed against the then prevailing concepts that excitation in the normal organism is always manifested as an "impulse," a momentary stimulus; in other words the two physiologists denounced the concepts which stated that excitation cannot be graduated or deepened, cannot become greater or smaller, but is absolutely identical and quantitatively constant for all excitable tissues.

The first entry in the physiological literature in support of the possibility of stationary excitation was made by Vvedenskii (1901-1903).

Modern electrophysiology has shown convincingly that excitation, in its physicochemical basis, is always a depolarizing process. Consequently, there is reason to suppose that where reversible depolarization takes place under the influence of certain transient influences, the conditions are right for excitation.

Without introducing the concept of stationary prolonged excitation into the physiology of the central nervous system it would be difficult to explain the physiological mechanisms of formation of the dominant focus and of temporary connections.

In my opinion, slow potentials are not produced by fusion or summation of individual waves of excitation but are a reflection of local prolonged activity. The stationary local potential rises or falls under the influence of the stimuli. It can be graduated. Slow changes of steady potential, recorded in some cases in the cortex, reflect stationary excitation. Slow potentials, like fast potentials, reflect a process of excitation, and present it in a different form, namely, as a slow stationary process. Just as fast changes of potential reflect a traveling wave of excitation, long-lasting changes of potential reflect slow, enduring changes of excitation. This "monistic" view of the relationship between excitation and its electrical component assumes the presence of foci of prolonged excitation, electrographically reflected by slow potentials, in the central nervous system as normal and extremely important factors in its function. Interest in slow potentials has thus increased particularly in recent years. This view also assumes that not only the action potential, but also the slow potentials reflect an active physiological state.

The theory of electrotonus, of the action of a steady current on the nervous system, is intimately connected with the problem under discussion.

In my investigations of slow cortical potentials I have attempted, first, to discover under what conditions slow changes of steady potential appear and, second, by applying a weak steady current to the cortex and certain subcortical structures, to simulate the natural change in the steady potential so as to create experimentally "foci of stationary excitation" possessing the properties of a dominant focus.

Slow cortical potentials in animals were investigated in our laboratory by Shvets (1958, 1960a, b, 1963), Grechushnikova (1963), Kuznetsova (1963) and Pavlygina (1964); enduring changes in membrane potential of the isolated nerve cell were studied by Ezrokhi (1967), and in the electromyogram by Rusinov and Chugunov (1938) and in the electrocardiogram by Borisova and Rusinov (1940).

In joint investigations with R. N. Lur'e (Lur'e, 1949; Rusinov, 1947, 1949) slow potentials were recorded, apparently for the first time, in the human EEG, predominantly in the frontal region. Slow potentials were seen particularly clearly during the formation of a CR to a combined stimulus (Lur'e and Rusinov, 1955). Similar slow potentials in the human EEG were described by Walter (1964); originally he called them "contingent negative variation," but later "E waves" (expectancy waves).

7. Slow Potentials in the Cortex during Defensive Conditioning

Marked changes in the steady potential level are observed during the formation of a temporary connection. As long ago as 1958 it was shown (Shvets, 1958; Rusinov, 1962) that during defensive conditioning in rabbits to photic stimulation, changes of SPL with an amplitude of several hundreds of microvolts or more, and a duration ranging from several seconds to several tens of seconds, can be recorded from the cortical surface during the combinations. The changes in general were predominantly in the negative direction. Also in 1958, Morrell (1962a) reported that the negative change in SPL arising in the cortex during electrical stimulation of the mesencephalic reticular formation can be reinforced as a CR by combining photic with this electrical stimulation. Later, results were published showing the existence of a conditioned neg-

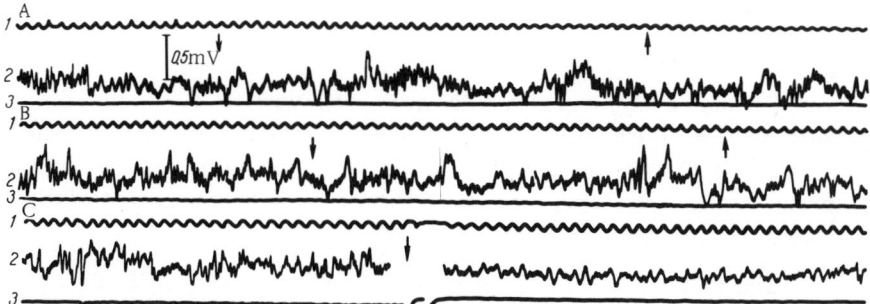

Fig. 5. EEG of the rabbit visual cortex; orienting reflex to photic and acoustic stimulation extinguished (work of Shvets). Photic (A), acoustic (B), and electrodermal stimulation (C) do not evoke a slow change of steady cortical potential. 1) respiration, 2) EEG, 3) movements. Time marker 1 sec; downward arrow indicates onset, upward arrow, end of stimulation. Nonpolarizable electrodes, dc amplification.

ative change in SPL in man (Walter, 1964, 1965; Irwin et al., 1966; Low et al., 1966; McAdam et al., 1966) and of conditioned changes, both negative and positive, in SPL in cats (Rowland et al., 1963).

Shvets (1960) showed experimentally that slow potentials are produced in the cortex during conditioning in rabbits. Electrical activity was recorded with the aid of a dc amplifier. Nonpolarizing ($Zn-ZnSO_4$) electrodes were used. The EEG was recorded from the visual and motor cortex on the right side through the thinned calvarium. A defensive CR to photic stimulation was formed. The electrodermal US was applied to the left forelimb. Usually at the beginning of the experiments the orienting reflex to the CS was extinguished. Depression of the background rhythm in response to the photic stimulus was well marked at first but disappeared by the 10th or 11th presentation. However, after combinations of photic and electrodermal stimulation, depression of the background rhythm to photic stimulation reappeared and continued in all subsequent experiments. Neither photic nor electrodermal stimulation, when applied separately, evoked a slow change of steady cortical potential in these experiments (Fig. 5).

On the appearance of the first CR, when the photic stimulus acquires the role of CS, slow potentials begin to be recorded in the visual cortex. Their amplitude ranges from 200 μV to several millivolts and their duration from a few seconds to several tens of seconds. On reinforcement with the US the slow wave in the visual cortex may be augmented. During subsequent combinations the change in steady potential in response to the CS also appears in the motor cortex.

Consequently, experiments on rabbits with dc recording of electrical activity with nonpolarizable electrodes showed that the formation of a temporary connection is accompanied by a change in the cortical steady potential in the form of a slow potential, appearing in response to the CS.

The same stimulus (photic), before combination with the US, did not evoke a change in the steady potential in either the visual or the motor cortex of the rabbit. Nor was such a change evoked by the US, electrodermal in this case.

According to Shvets, fluctuations of SPL in response to any afferent stimulus during conditioning are observed in the generalization phase of the CR. In this phase an indifferent stimulus, to which the orienting reflex had previously been extinguished, as shown by desynchronization of electrical activity, like the CS evokes a change in SPL (Figs. 6 and 7). The characteristic feature of the potential in these cases is the presence of slow waves in the aftereffect also.

A slow potential on the cortical surface is produced not only during conditioning. Traumatic dominants were investigated by Ryabinina (1959). Gentle pressure was exerted on the motor cortex of a rabbit, in the zone of representation of the forelimb, through a small hole in the skull by means of a special device. If, under these conditions, indifferent stimuli (acoustic, photic),

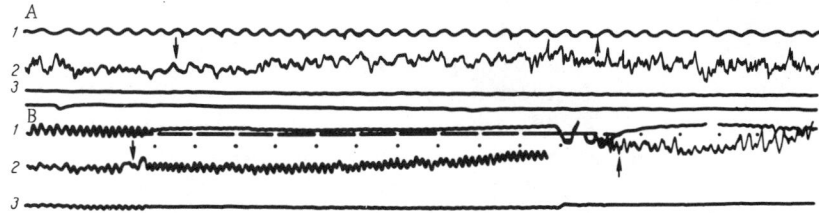

Fig. 6. EEG of the visual cortex of the same rabbit as in Fig. 5. Generalization phase during defensive conditioning (after Shvets). Indifferent stimulus (A) and photic stimulus (B) evoke a change of steady cortical potential. Upward deflection corresponds to negativity. Remainder of legend as in Fig. 5.

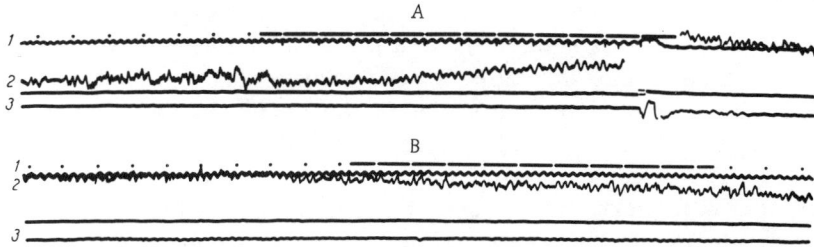

Fig. 7. Slow negative change of steady cortical potential appears in a rabbit during defensive conditioning in response to the CS (photic, A) but not to the indifferent stimulus (acoustic, B) (after Shvets). Top line: marker of time (in sec) and stimulation. Remainder of legend as in Fig. 5.

to which the orienting reflex had first been extinguished, were applied a motor response of the corresponding forelimb developed to these stimuli. In other words, a zone of mild trauma in the cortex can become a dominant focus.

Under the same experimental conditions Shvets (1960a) investigated changes in electrical activity in a dominant focus induced by gentle pressure. She found that at a certain level of negativity in the dominant focus, hitherto indifferent stimuli began to evoke slow potentials accompanied by movement of the corresponding limb. Before pressure on the motor area, the same stimuli produced neither movement of the forelimb nor a long change of cortical SPL. A change in cortical SPL was necessary before these indifferent stimuli could give rise to long, slow fluctuations of potential.

The results of these experiments indicate that long aperiodic slow changes of SPL on the cortical surface are not themselves specific for CR formation. They reflect the presence of stationary excitation with its fluctuations of a prolonged character. Where the conditions provided favor the onset of stationary excitation and a certain level of negativity (a dominant focus, foci during conditioning, prolonged stimulation in exposed areas of cortex), clear fluctuations of SPL appear in response to afferent stimuli. Under those conditions they may also appear "spontaneously" in response to afferent stimulation from the external or internal environment disregarded by the experimenter. Kogan (1949) observed slow waves at 2-5/sec during conditioning.

Slow potentials (changes in SPL), while not specific for a temporary connection, nevertheless exhibit characteristic differences in that case. Shvets studied the spatial distribution of slow potentials over the cortex during defensive conditioning to a photic stimulus. She implanted 10 electrodes in the region of the motor cortex and the same number of electrodes in the region of the visual cortex. The silver or platinum electrodes rested on the dura, and were 1-2 mm apart. They were 40μ in diameter and insulated except at the tip. In these experiments Shvets was only able to show that slow waves of steady potential were present during conditioning but she could not determine the absolute magnitude of the potential. She found an interesting phenomenon: strict localization of the slow potentials. These waves, recorded during conditioning from a series of electrodes, originally in the visual cortex and later in the motor cortex, are recorded from progressively fewer electrodes as stabilization of the temporary connection takes place and the number of combinations is reduced, and ultimately they are recorded from only one electrode.

In her later investigations Shvets showed that during conditioning the changes in SPL are altered not only in their cortical distribution, but also in their direction. Before the combinations, and after extinction of the orienting reflex to it, a photic stimulus does not evoke changes in SPL. The first changes in response to the photic stimulus during the combinations appeared as a rule in the visual cortex as negativity. Later, however, changes in SPL spread to all the cortical recording points (up to 10), and became positive in sign. With an increase in the number of combinations the SPL, although still remaining diffuse, once again became negative and lost its diffuse character only gradually on becoming localized in the motor and visual areas.

Concentration of slow potentials in the cortex during stabilization of the temporary connection is a characteristic feature of their manifestation during conditioning. Another characteristic feature, revealed also in experiments with nonpolarizing electrodes, is their strict "conditionality." Slow potentials recorded in the generalization phase in response to all stimuli applied, once the temporary connection is stabilized and differentiation established, appear in the cortex only in response to the CS, but not to the differential stimulus.

What is the functional role of changes in SPL recorded from the cortical surface? Changes in SPL during conditioning can be considered to reflect the active state of neurons in the investigated cortical areas, for CRs take place against the background of SPL changes. Other evidence is the gradual decrease and disappearance of local negative changes in SPL during extinction of the CR and the formation of differentiation. It is interesting to note that under these conditions the changes in SPL simply disappeared and were not replaced by changes of a different character reflecting an "inhibitory state" as might have been expected. The evident absence of SPL changes may be due to strong concentrations or equal magnitude of the fields of opposite sign arising on the surface.

8. Long-Lasting Depolarization of the Single Neuron

The investigator is interested not only in the presence of a particular bioelectrical process and its relationship to the functional state of its substrate, but also in its role in functional links between neurons. By analogy with other bioelectrical phenomena (action potentials, muscle end-plate potentials, synaptic potentials, and so on), it would be expected that steady potentials not only reflect changes in the degree of polarization of the membrane, but also actively change the unit activity in the central nervous system. The old dispute about this issue falls under two headings.

Hardly anyone denies that when the membrane potential of part of a cell changes, the excitability of its other parts also changes. Opinions differ only as regards how it changes.

A more debatable question is the role of external electrical fields in the activity of the central nervous system, i.e., to what extent a passive change in membrane potential can change unit activity. Rusinov (1951, 1953a, b) postulated the leading role of stationary fields, i.e., of slowly changing potentials, in the closing of temporary connections. This hypothesis served as the basis for laboratory investigations which showed that anodal polarization of the rabbit sensorimotor cortex by a weak direct current (of density less than 1 $\mu A/mm^2$) at the point of representation of one of the animal's limbs has the result that previously indifferent stimuli (photic, acoustic) begin to evoke contractions of the corresponding limb. The main conclusions drawn from these investigations will be examined below.

The question of the nature of slow electrical potentials and the possibility of interaction between neurons through the extracellular electrical field which they generate were investigated by Ezrokhi (1967, 1968, 1969, 1970), working in this laboratory, on nerve cells of the isolated crustacean stretch receptor.

The isolated preparation consists of two receptor muscle bundles attached to the shell and two neurons whose dendrites are in close contact with the corresponding muscle bundle. Axons of each neuron after a distance of about 1 mm join to form a nerve trunk which also includes a number of motor fibers running to the muscle

8. LONG-LASTING DEPOLARIZATION OF THE SINGLE NEURON

bundles and an inhibitory fiber terminating in synapses on the distal branches of the dendrites of both neurons. The soma is about $50\,\mu$ in diameter (Fig. 8).

These receptor neurons are similar to many types of nerve cells of the central nervous system not only in their anatomical structure but also in their basic physiological characteristics

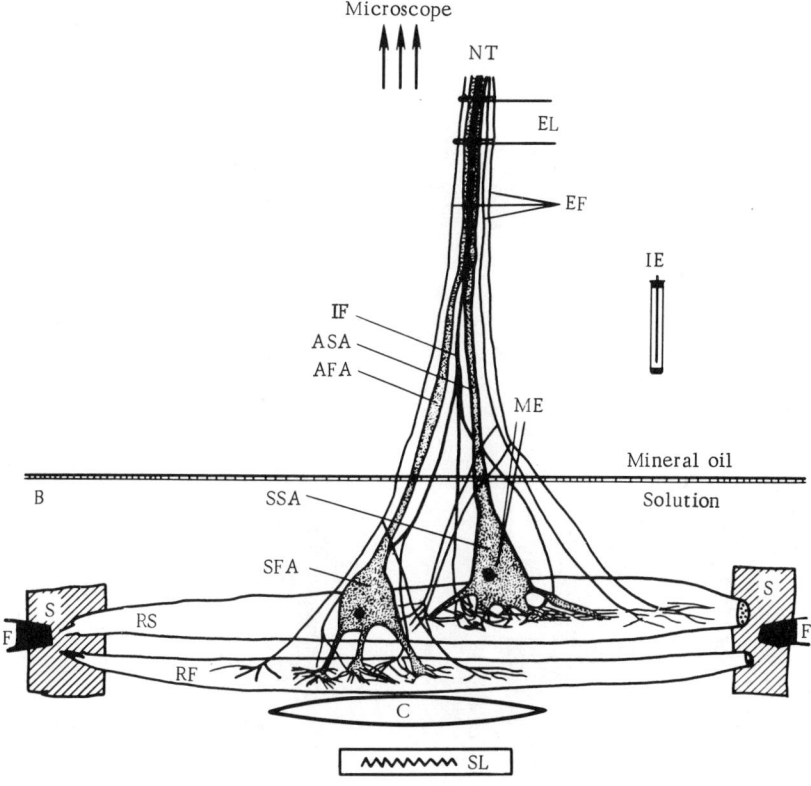

Fig. 8. Diagram of isolated crustacean stretch receptor preparation and its position in Ezrokhi's experiments. NT) nerve trunk; IF) inhibitory fiber; ASA) axon of slowly adapting (SA) neuron; AFA) axon of fast adapting (FA) neuron; SSA) soma of (SA) neuron; SFA) some of (FA) neuron; RF) receptor muscle of (FA) neuron; RS) receptor muscle of (SA) neuron; EF) efferent fiber; S) shell; F) forceps; IE) indifferent electrode; ME) microelectrode; EL) electrodes on nerve trunk; B) boundary between mineral oil and solution; SL) source of light; C) condenser.

CHAPTER II: THE PROBLEM OF STATIONARY EXCITATION

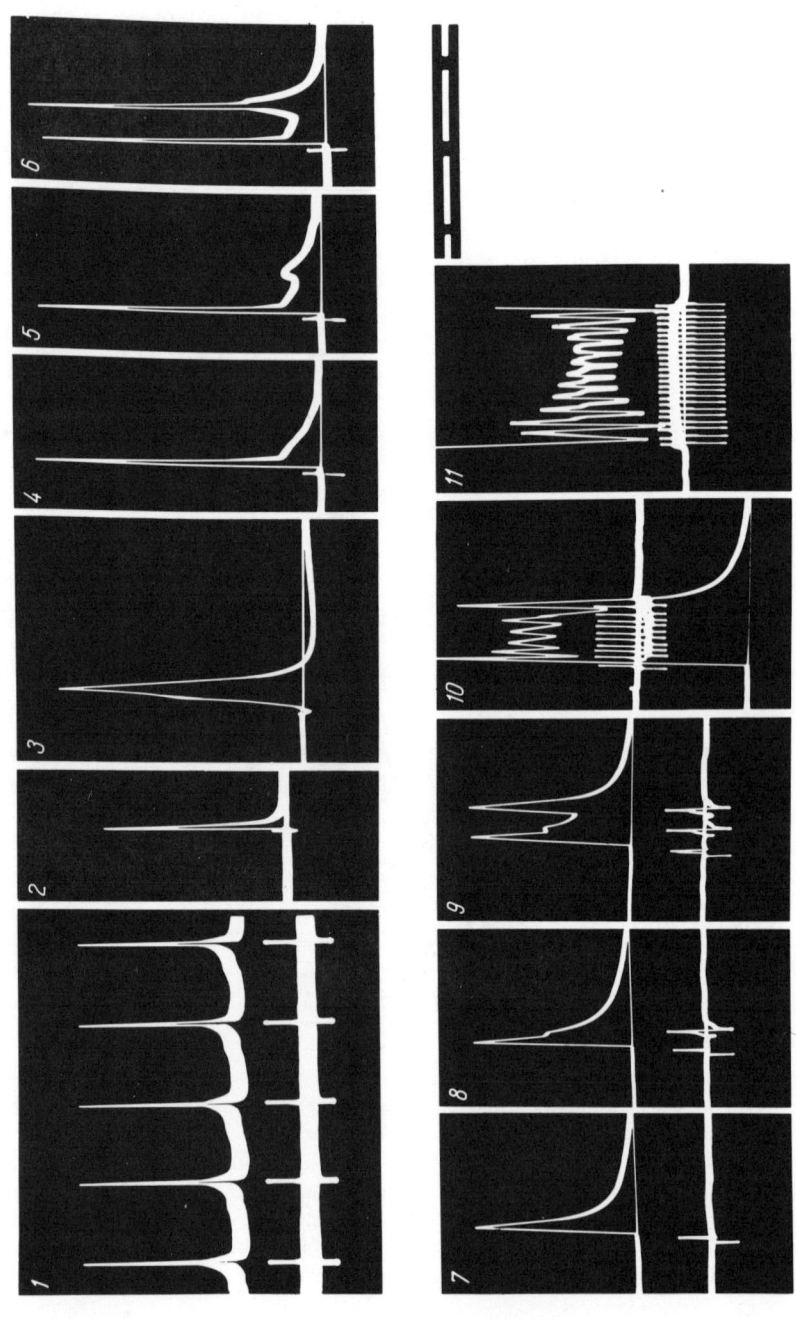

Fig. 9. Activity of normal and altered neuron in response to orthodromic and antidromic excitation (work of Ezrokhi): 1) repetitive discharges of soma (above, intracellular recording) and axon in response to stretching muscle bundle; 2,3) intracellular action potentials of soma in response to antidromic excitation (components of initial segment and soma-dendritic region are seen); 4-6) states of appearance of second AP in soma in response to antidromic excitation of damaged neuron; 7-11) stages of formation of grouped discharges after application of strychnine to orthodromically excited neuron; number of APs on axon (8-11, below) is greater than in soma. Calibration for intracellular recording (in mV): 44 (1, 2, 7-9); 22 (3-6, 10), 11 (11). Time calibration (in msec): 41 (1, 2, 10, 11), 20 (4-9), and 8.5 (3) between beginning of two dotted lines.

(resting potential, action potential and site of its generation, critical level of depolarization, impedance and time constant of the membrane). One of the two neurons is slowly adapting (SA), the other fast adapting (FA), so that after sudden stretching of the muscle bundles the action potentials (APs) generated by the FA neuron quickly decrease in frequency and cease altogether within a few seconds, whereas the APs of the SA neuron continue to appear regularly, in accordance with the degree of stretching, remaining almost unchanged in frequency for several hours.

The preparation, held by the shell to which the muscle bundles are attached, was immersed in van Harreveld's solution and the nerve trunk lifted on two silver electrodes in the layer of mineral oil by means of micromanipulators. Intracellular recordings were taken from the soma of the SA neutron. The grid current of the cathode follower did not exceed 2×10^{-12} A after working for 5 h. The nonpolarizable (Ag–AgCl in agar, made up in physiological saline) microelectrode was filled with 3 M KCl or 0.6 M K_2SO_4 solution and its resistance was 20-30 MΩ. The indifferent nonpolarizable electrode was placed in physiological saline a few centimeters away from the preparation. Recordings were made on a dual-beam Disa CRO, one input of which was connected to the output of a cathode-follower and the other, through a preamplifier with symmetrical input, to the electrodes on the nerve. Bipolar recordings were taken from the axons, with interelectrode distance 2-4 mm. Intracellular polarization was applied from the radiofrequency output of a Physiovar stimulator using the bridge circuit of Frank and Fuortes. Strychnine or procaine for testing

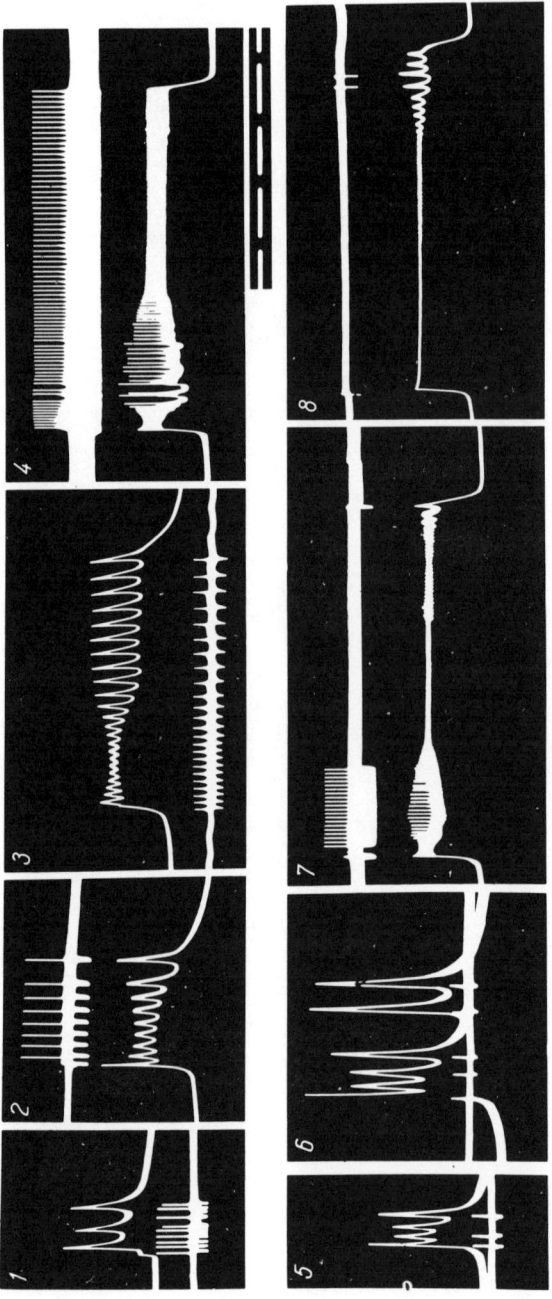

Fig. 10. Correlation between somatic and axonal action potentials during grouped discharges (work of Ezrokhi): 1, 2, 4-8) grouped discharges evoked by procaine; 3) by strychnine; 1) number of APs in axon is 3 times greater than in soma; 2-4) number of APs in axon and soma is the same; 5-8) number of APs in axon smaller than in soma; 2, 4, 7, 8) recording from same neuron; 1, 3, 5) recording from axon (below). Calibration for intracellular recording (in mV): 44 (1-5, 7, 8), 22 (6). Time calibration (in msec): 41 (1-3, 5, 6), 200 (4, 7, 8).

were added to the surrounding solution close to the preparation from a syringe.

Action potentials (APs) arising in response to moderate stretching of the muscles are shown in Fig. 9. The resting potential of the membrane was about 70 mV, the amplitude of the AP 80 mV, its duration 1.8-2 msec, and the critical level of depolarization 12 mV. Each AP is preceded by a gradually increasing generator potential, changing into a prepotential. The AP terminates with a positive deflection which returns the membrane potential to the level of the resting potential. Slight (up to 5 mV) depolarization of the membrane takes place 1-5 min after addition of strychnine or procaine to the solution, and it is accompanied by an increase in the frequency of the APs or by their appearance if the original stretching was subthreshold. During the further action of strychnine or procaine, grouped discharges and long action potential (LAPs) are formed. Each discharge consists of an AP of high amplitude and with a rapidly rising phase of depolarization, in which the phase of repolarization is interrupted by the next AP, of lower amplitude. The first AP changes into a plateau of depolarization, on which are superposed waves of different frequency, amplitude, and duration, closely connected with the degree of negativity of the soma. These grouped discharges are extremely varied in appearance. Grouped discharges of each neuron may vary depending on the degree of action of the strychnine or procaine. With the highest level of depolarization these waves may become small or disappear altogether. In other cases depolarization is greatest immediately after the first AP, after which the negativity is reduced and waves appear on the plateau. The subdivision into grouped discharges and LAPs is conventional and it simply reflects the rate of increase and the magnitude of depolarization of the membrane.

Correlation between the APs of the soma and axon during grouped discharges is shown in Fig. 10. In some cases the number of axonal APs is several times greater than the number of somatic (Fig. 10), and it may reach 400/sec. Specimens of grouped discharges in which the number of spikes is the same in axon and soma are shown in Fig. 10, 2-4. Clearly in the recording from the soma there are no additional nonspike waves. Finally, in some grouped discharges the number of APs in the soma is greater than the number of APs in the axon (Fig. 10, 5-8). As these examples

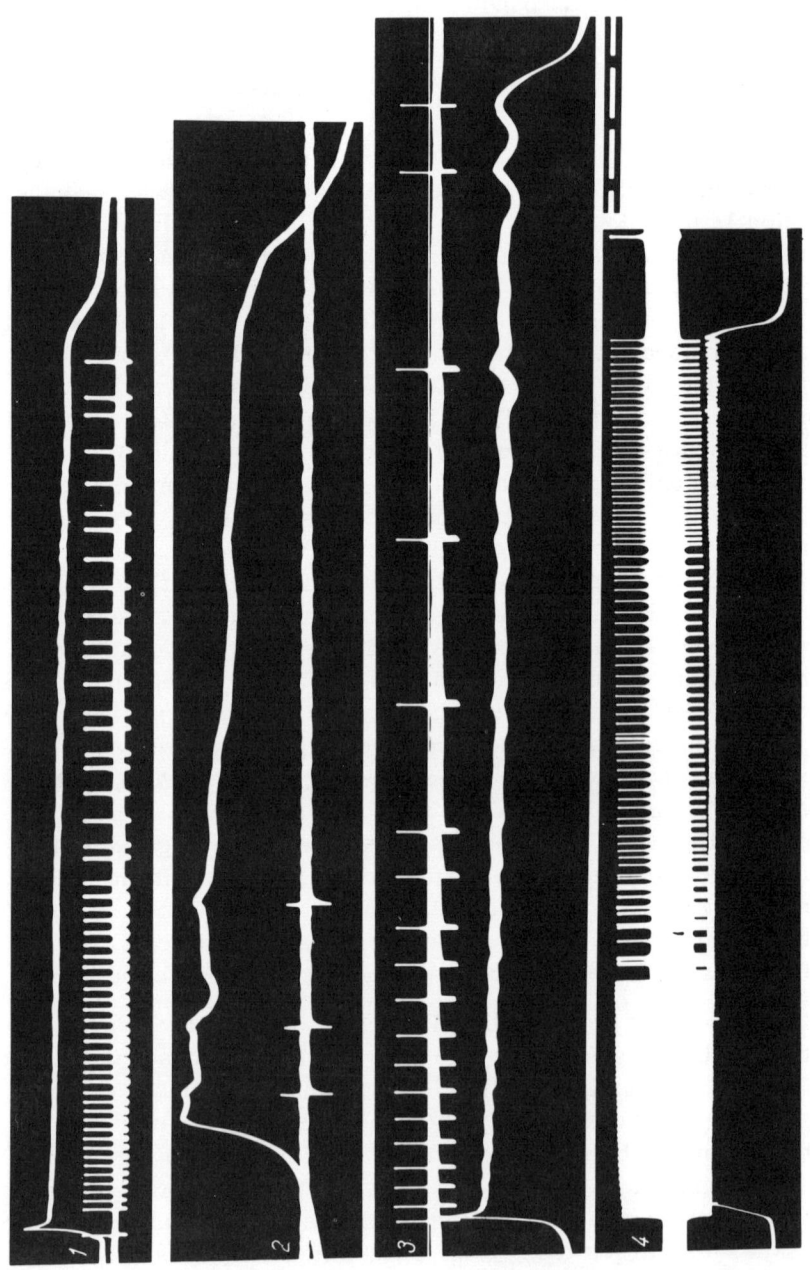

Fig. 11. Action potentials in axon during strong depolarization of soma evoking grouped discharges (work of Ezrokhi): 1, 3) grouped discharges evoked by procaine; 2, 4) by strychnine; 2) 1st AP of grouped discharge in soma (above) is not accompanied by an axonal AP; 3) some waves of membrane potential in soma evoke axonal AP; 4) clear correlation between action potentials of axon and oscillations of soma is observed at beginning of grouped discharge and at its end, when amplitude of somatic oscillations rises (below). Calibration for intracellular recording (in mV): 44 (1, 4), 22 (2, 3). Time calibration (in msec): 41 (1-3), 200 (4).

---

of the various types of grouped discharges (Fig. 10, 5-8) show, there are no significant qualitative differences between the somatic APs whether or not they are accompanied by axonal APs. Those which are not so accompanied are lower in amplitude.

If the dose of strychnine or procaine was large enough, the gap between neighboring groups became less clear, and finally the record had the appearance of continuous, slow, low-amplitude waves against the background of continued depolarization of the membrane (Fig. 11). Depending on the dose of strychnine or procaine applied, these transitional stages varied from a few hours to a few seconds in duration.

Because of diffusion of the added strychnine or procaine from the site of its introduction near the preparation into the surrounding solution, or by changing the solution, normal activity can be restored at any stage. Moreover, this procedure can be repeated several times. Temporary recovery can be achieved by hyperpolarizing antidromic excitation or by application of a direct pulse of current of sufficient strength and duration. The action of strychnine and procaine gave identical results although the effect of procaine was slightly stronger in the same concentration.

These results thus show that the action of strychnine or procaine on the receptor neuron converts regular APs into grouped discharges and LAPs against the background of an only slightly reduced (by 5 mV) SPL, and that if the action of these substances is prolonged, progressive depolarization of the membrane results. The grouped discharges and SPL are regularly converted into lasting depolarization of the membrane, accompanied by small fluctuations of potential.

The transition from depolarizing waves with discharges on them, but with an unchanged SPL, to lasting depolarization accompanied by small oscillations was described by Sawa, Usuni, and Kaji (1965) after application of strychnine to cortical neurons. In all investigations with intracellular recordings from cortical neurons of a focus produced by a variety of chemical agents, the presence of depolarizing waves with discharges on them is the most characteristic property (Li, 1959; Matsumoto and Ajmone-Marsan, 1964).

There are two possible explanations of the mechanism of this cortical unit activity. Most investigators explain the depolarizing waves by the summation of EPSPs, others by the internal properties of the cell membrane itself. Since nerve cells of crustacean stretch receptors have no excitatory synapses, it is evident that the response of the receptor neuron, when treated with strychnine or procaine, is based on changes in the membrane properties of the neuron itself. It thus follows that, although the possible role of postsynaptic potentials in the long depolarization of cortical neurons cannot be ruled out, the transition from separate APs to grouped discharges, LAPs, and even to lasting depolarization can be explained by a change in the membrane properties of these neurons.

The nature of the epileptiform discharges in the aftereffect, evoked by sufficiently strong electrical stimulation of the cortical surface and persisting for several seconds after removal of the stimulus, is interesting.

The global epileptiform surface waves in an experimental epileptic focus are known to correspond to waves recorded intracellularly, namely, to the paroxysmal depolarization changes (PDCs). Just as with the LAPs two hypotheses have been put forward to explain the neurophysiological basis of the PDC. According to one of them, the original cause of PDC generation is a change in spike electrogenesis, a change in the properties of the membrane itself. According to the other, PDCs are extremely potentiated EPSPs.

Voronin (1970) investigated the action of local polarization on afterdischarges of cortical neurons of the unanesthetized rabbit, but could not confirm the hypothesis that PDCs arise through a disturbance of spike electrogenesis. His observations are evi-

8. LONG-LASTING DEPOLARIZATION OF THE SINGLE NEURON

dence rather that the PDCs are modified, extremely potentiated, EPSPs. "Nevertheless," he writes, "the results strictly speaking do not prove the synaptic nature of the PDC and they can be interpreted within the framework of the membrane hypothesis. Comparison of PDCs and LAPs shows that the gradual character of the change in PDC in some cases, the shape of the PDCs, their appearance in depolarized neurons with an inactivated mechanism of action potential generation — these cannot be regarded as confirmation of the synaptic nature of PDCs, as some have proposed, because LAPs can possess these same properties."

Grouped discharges and LAPs have also been observed in the soma of the receptor (Washizu, 1965) and other neurons (Gerasimov, Kostyuk, and Maiskii, 1964), and also in muscle and nerve fibers on elevation of the temperature, a change in the composition of the intracellular and external medium, and so on.

Stable depolarization has been observed in the crustacean receptor neuron by Washizu (1965). However, he also found that the initial level of membrane potential is restored only by application of a direct anodal pulse. Similar stable depolarization and its restoration by a hyperpolarizing current were described by Tasaki (1959) in neurons of spinal ganglia treated with Ba^{++} ions. Grouped discharges, LAPs, and stable depolarization are thus a general response of neurons and muscle fibers to nonspecific altering agents.

In some cases, during the action of strychnine or procaine, APs which do not correspond to activity recorded intracellularly from an SA neuron are recorded from electrodes placed on the nerve trunk. These potentials have a higher amplitude than impulses traveling along the axon of the SA neuron, and under normal conditions at the beginning of the experiment they arise during stretching and during the first few seconds thereafter, and they are thus APs of a SA neuron. The negative-positive phase sequence indicates that the APs spread to the electrodes away from the soma. In most cases activity of the SA neuron disappeared a few seconds after its appearance, but sometimes it persisted for hours, throughout the period of action of the agents applied. The neuron ceases to be fast-adapting and it gives rise to APs with a high degree of regularity. Activity of the SA neuron undergoes the same changes under the action of strychnine or procaine as

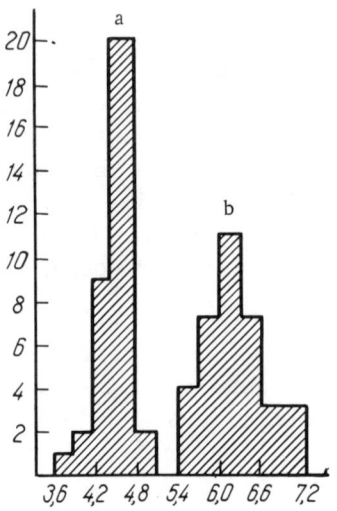

Fig. 12. Histograms of activity of FA neuron during LAP plateau (a) of SA neuron and in intervals between them (b) (work of Ezrokhi). Abscissa, arithmetic mean of time intervals between APs of FA neuron in conventional units; ordinate, number of successive LAPs of SA neuron or of intervals between them.

the activity of the SA neuron, but with considerable delay, so that while the SA neuron responds with an LAP, the FA neuron discharges single regular APs.

Ezrokhi's experiments show that the spike frequency of the FA neuron during the LAP of the SA neuron is higher than in the period between LAPs. In different preparations, the increase in activity of the SA neuron during the LAP plateau of the SA neuron varies. In some preparations, for instance, in which the difference was statistically significant, the difference between the arithmetical mean values was about 4%, while in others it reached 25%. The difference was statistically significant in certain experiments. In each of them it was in fact highly significant ($P < 0.001$). Frequencies of the action potentials of a FA neuron were compared in intervals between successive (10-40) LAPs of the SA neuron and during the LAP plateau. Histograms of one experiment are shown in Fig. 12. Sometimes the LAP plateau of the SA neuron converts the single AP of the FA neuron into a double AP.

Regularly discharging neurons are highly sensitive to changes in the level of membrane potential. Depolarization of the soma membrane by a current of 0.17×10^{-9} A, for instance, increased the frequency by 50%. The inward resistance of the membrane, measured by imbalance of the bridge on withdrawing the microelectrode from the soma, was 3.1 MΩ. The applied current thus depolarized the membrane by 0.5 mV. Even depolarization of the cell membrane by 0.1 mV led to an appreciable (by 5%) increase in activity. Meanwhile, to cause the appearance of an AP in a neuron not discharging spontaneously, much greater (several millivolts) depolarization of the membrane was required. After removal of the depolarizing current, activity of the neuron ceases or becomes slower than its firing rate before application of the current. It gradually reaches its initial level. The strength and duration of inhibition after cessation of the depolarizing current depend on the strength and duration of this current. In many cases, delay in the appearance of the AP or LAP of the FA neuron was observed even after the steep decline of the plateau of the SA neuron.

These results show that a functional link can be created between two receptor neurons. This link appears during LAP formation in the SA neuron, and it is expressed primarily as an increase in the AP frequency of the FA neuron during the LAP plateau of the SA neuron. How is this link brought about? Stretching one receptor muscle bundle may lead to stretching of the second bundle through the presence of connective tissue between them. However, the triggering of the LAP does not result from movement of the muscle bundle, as the following facts confirm: 1) the regular discharges of LAPs are formed gradually from regular series of APs and, just as during normal activity they have a phase of gradual growth of generator potential; 2) during the LAP sequence, microscopic examination revealed no muscle activity; 3) activity of the FA neuron is increased exactly from the beginning of the plateau of the SA neuron and it continues just as long as this plateau exists; the LAP evoked by a transient intracellular pulse of current activates the FA neuron.

The next possibility is the existence of a morphological substrate for transmission of, it is claimed, electrotonic effects. The link between neurons of the lobster cardiac ganglion, the mollusc abdominal ganglion, the giant cells of the segmental ganglion of the leech, and so on, is attributed to the hypothetical existence of bridges (Hagiwara and Morita, 1962; Bennett et al., 1963, 1967a, b, c, d, etc.). On the other hand, investigations of some situations have revealed synapses through which transmission takes place electrically (Furukawa et al., 1963). Microanatomical investigations of the crustacean stretch receptor (Florey and Florey, 1955; Peterson and Pepe, 1961) have yielded no evidence in support of the existence of a bridge between neurons. The most likely explanation of this effect of activity of the SA neuron on activity of the FA neuron is an electric field generated during high-amplitude depolarization of the membrane associated with LAP generation. The high sensitivity of the neuron, discharging with spontaneous regular APs, to the voltage gradient developing in the surrounding solution would also contribute to this result. A full quantitative treatment of sensitivity of the SA neuron of the crustacean receptor to the voltage gradient is given in the paper by Terzuolo and Bullock (1956).

The results of Ezrokhi's experiments show that there is a real possibility of a physiologically significant effect, which must include synchronization of activity, to be produced by an extracellular electric field generated by neurons of the central nervous system. The sensitivity of neurons of the central nervous system to AP modulation by an external field, according to the available evidence, may actually be higher than for neurons of the crustacean stretch receptors (Strumwasser and Rosenthal, 1960). Additional factors are the extremely close packing of the neurons (Karlson and Schultz, 1964) and their architectonics.

These results, indicating an increase in the activity of one neuron in response to a long, high-amplitude shift of membrane potential of another neuron, thus point to a possible role of the extracellular electric field generated by the nerve cells themselves in the functional interneuronal connection.

* * *

Slow electrical potentials in the cortex have many different names in the literature. They have been recognized in the world literature only comparatively recently, in the last 10-15 years. Steady potentials in the central nervous system are an equally widespread bioelectrical phenomenon as electrical fluctuations in other parameters, or perhaps even more widespread.

8. LONG-LASTING DEPOLARIZATION OF THE SINGLE NEURON

In response to a flow of afferent impulses, fields of steady potential, varying in both origin and direction, are generated in the projection area and also in other parts of the cortex. In response to relatively weak stimuli positive fields are predominant and most changes are observed in the specific region; with an increase in stimulus strength, negative changes become predominant and most of the changes take place in the anterior regions.

Marked changes in the level of the steady cortical potential are observed during formation of a temporary connection. Slow waves of cortical potential recorded in the generalization phase to all stimuli applied, the orienting reflex to which had previously been extinguished, appear in the cortex during stabilization of the temporary connection only in response to the CS. They do not appear in response to a stimulus to which differentiation has been formed. Concentration of slow potentials (changes in SPL) in the cortex during stabilization of a temporary connection is a characteristic feature of their behavior during conditioning. Changes in cortical SPL during formation of the temporary connection reflects the active state of the cortical point recorded; the CR takes place against their background.

Intracellular recording of potentials on single units not possessing excitable synapses (the crustacean stretch receptor) has shown that the transition from single APs to grouped discharges, LAPs, and from thence to long-lasting depolarization can be explained by changes in the membrane properties of the neurons themselves.

Chapter III

A Model of the Cortical Dominant Focus Produced by Weak, Steady Current

1. Polarization of the Rabbit Motor Cortex by Steady Current

One of the problems facing electrophysiological investigation of the central nervous system is how to determine the form of the functional link between one neuron and another. What types of excitation are responsible for interneuronal connections both in unconditioned-reflex activity and during formation of temporary connections?

One feature which distinguishes the higher levels of the central nervous system is that, unlike at lower levels, regular electrical activity is continuously observed. In the nerve trunk and spinal cord, electrical phenomena in the form of single and repetitive waves arise only during reflex activity, when effects are present in a working tissue or in the period immediately after them. In the medulla, however, electrical waves are observed even in a relatively resting state, with no special stimulation of afferent systems. A particularly clear, regular rhythm, in the form of a dominant alpha-rhythm (8-12/sec) is observed in the human cortex in a relatively quiet, waking state. The electrical effect of excitation and the form of the potential depend on the structure which is excited and not on its functional state.

The rhythm of electrical activity recorded in the EEG indicates primarily a functional link between neurons (Rusinov, 1951, 1953a, b). The EEG and its special features depend primarily on the character of the interneuronal link between the structure whose EEG is recorded and on its functional state at the time of recording.

What type of interneuronal communication plays the essential role in closure of the temporary connection? Considering that local activity, if expressed as alpha-waves, as slow pathological waves, or as waves of another, approximately regular type, is formed from slow potentials, and that the most adequate reflection of the "functional state" of a system of neurons is the level of their steady potential, there appeared to be a good case for the following hypothesis: the type of interneuronal communication in the form of electrotonic influences plays the most important role in the formation of temporary connections (Rusinov, 1951, 1953a, b). Hence it follows that extracellular fields, exerting electrical effects on systems of neurons, can actively create and maintain a state of unison between two distant cortical points, and, it must be supposed, a united response to stimulation, so that a stimulus previously neutral as regards a particular activity can become an adequate CS and can replace the US.

Communication in the form of an electrotonic influence is closest to the action of a steady current. Consequently, if the sensorimotor cortex of a rabbit is stimulated, in the zone of representation of the forelimb, by a steady current from a focal electrode (the indifferent electrode being on the ear), which does not itself evoke any effect at the periphery, but merely changes the state of the particular cortical area concerned, and if during this time another (for example, photic) stimulus is applied, if this hypothesis is true a temporary connection will be formed between the two foci of excitation, and the limb will contract in response to photic stimulation.

The first experiments in this direction and their results were described by Novikova, Rusinov and Semiokhina (1952). A focus of lasting excitation was produced by the anode of a direct current in the zone of representation of the left forelimb in the sensorimotor cortex of an adult, unanesthetized rabbit. The focus of ex-

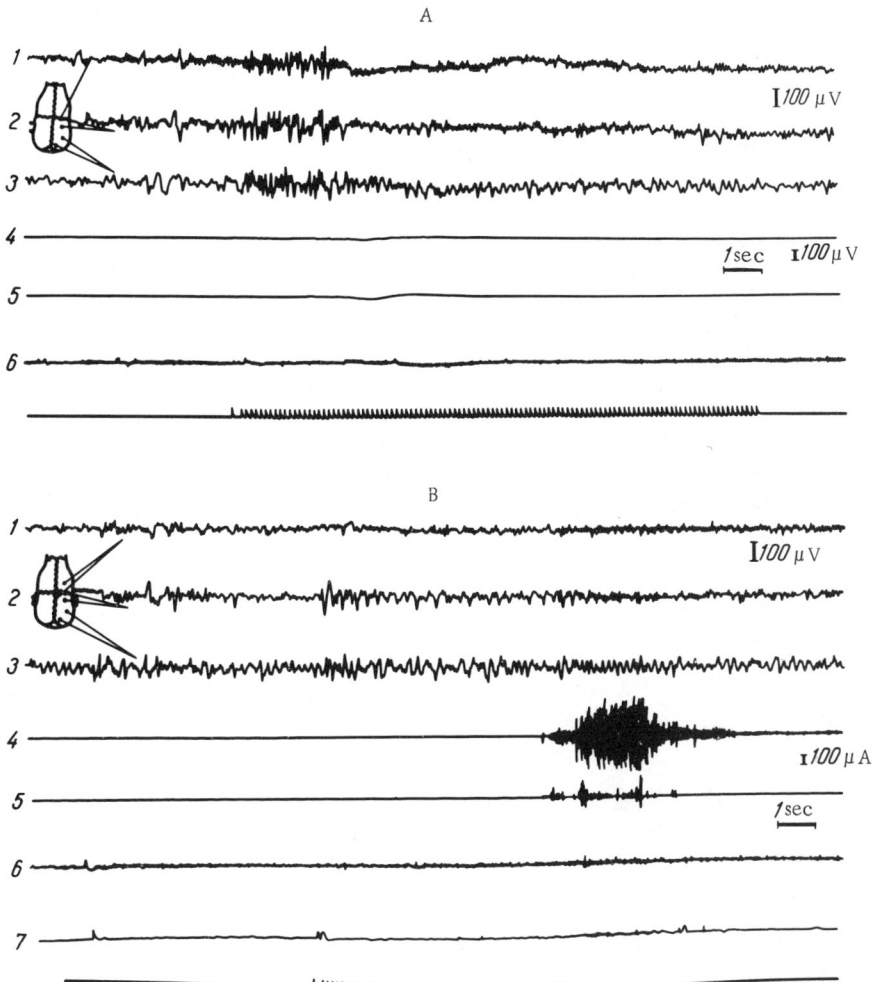

Fig. 13. Formation of a focus of excitation by polarization with the anode of a weak direct current in the cortical area for the left forelimb (work of Sokolova). A) Rabbit's motor response to regular photic stimulation is absent before dc polarization; B) motor response mainly of the left forelimb to similar regular photic stimulation during dc polarization (1.5 μV); 1, 2, 3) EEGs corresponding to diagram; 4) electromyogram (EMG) of left forelimb; 5) EMG of right forelimb; 6) EMG of left hind limb; 7) EMG of right hind limb. Bottom line marker of photic stimulation (8/sec).

citation was localized in the center for the forelimb, and not the hind limb, because of the fact, first described by Bekhterev (1906) and subsequently confirmed in experiments with steady current, that the projection zone of the forelimb in rabbits is relatively more precisely localized in the cortex than the projection zone of the hind limb. Polarization of the brain began with a weak steady current, the strength of which was gradually increased from 0.1 to 10 μA. During the action of the steady current, acoustic or photic stimulation was applied. Electrical activity of the cortex and limb muscles during the steady current and after its removal, and changes in the EEG associated with acoustic and photic stimulation were recorded. After the first orienting reflexes the rabbits ceased to give a motor response to subsequent stimuli or, in some cases, they responded by a slight and uniform contraction of all the limbs. In each experiment the purest "background," i.e., complete absence of a motor response to acoustic and photic stimulation, was created by repeated stimulation and extinction of the orienting reflex.

Under the influence of the anode of the steady current the activity in the sensorimotor cortex changes. Slow waves and fast fluctuations of potential appear in the EEG. Changes in brain activity of this type are not observed under another pair of recording electrodes, thus demonstrating the local character of the action of the weak steady current.

The focus of excitation produced by the action of steady current on the cortex possesses dominant properties. Afferent stimulation (photic, acoustic) against the background of the steady current gives rise to movement of the corresponding limb (Fig. 13).

The formation of this type of temporary connection between the visual and motor systems when a focus of excitation is present in the sensorimotor cortex can be observed dozens of times in the course of the experiment.

Analysis of the mechanism of the dominant reveals the following important fact. The focus of excitation produced in cortical polarization experiments begins to exhibit its dominant properties and to summate excitation from extraneous stimuli only if the steady current is of a certain strength. If variations took place during the short-term experiments on rabbits, the optimum current for dominant formation was found to be 1-2.5 μA (optimum current density 0.8-1 μA/mm^2) or higher. In long-term experiments with implanted electrodes, this optimum was lower still. If the strength of the current is above optimum, an inhibitory effect is observed: neither acoustic nor photic stimuli evoke contraction of the limb muscles.

It is only necessary to reduce the strength of the current, and thereby to bring the level of excitation in the sensorimotor area to its previous value for the region of dc polarization to resume its dominant properties. In the same experiment, by changing the strength of the direct current arbitrarily, a dominant can be induced and converted into an inhibitory state several times. Inhibition of a dominant can be produced not merely by changing the strength of the direct current, but simply by increasing the duration of its action on the cortex. The phenomena of "optimum" and "pessimum" (to use Vvedenskii's term), and conversion of a dominant focus into inhibition can also be obtained during a rapid sequence of afferent stimuli. Increasing the interval between two stimuli is sufficient for them to begin to evoke contraction of the limb muscles once more.

Whereas too frequent stimulation converts the dominant focus to inhibition, the dominant itself is not produced if the afferent stimulation is too weak. In other words, the temporary connection between the dominant focus and a particular sensory system is formed only at a certain level of excitation of that system. That is why a certain threshold strength of acoustic and photic stimulation is required to evoke movement of the limb.

For the same reason, sometimes the dominant focus is reinforced by acoustic, but not by photic stimulation in these experiments, but by choosing the appropriate optimal strength of stimulation the dominant focus can be reinforced by both sensory systems.

It is only under optimal conditions that reinforcement of the dominant focus by acoustic or photic stimuli in short-term experiments leads to isolated movements of the corresponding contralateral limb only. Most frequently afferent stimulation evokes movement of both forelimbs. However, contraction of the muscles of the contralateral forelimb are definitely predominant.

The dominant focus produced in the sensorimotor cortex is maintained throughout the action of a steady current for 4-6 h. During this period nearly every acoustic or photic stimulus evokes limb movement. It is in such cases of a lasting dominant, clearly manifested throughout the experiment, that the range of optimal strength of the direct current is widened. Conversely, if the dominant focus does not arise immediately and if it lasts for only a short time, the range of optimal strength of the direct current is very narrow; as soon as the current is increased a little the dominant is inhibited.

Experiments in which a focus of excitation is reinforced after removal of the direct current, i.e., in the afterperiod of its action, are particularly interesting in connection with inertia of the dominant. The time during which a focus of excitation in the sensorimotor cortex preserves its dominant properties after removal of the direct current varies in short-term experiments from a few to 20-30 minutes.

Some interesting results on inertia of the dominant focus have been obtained in experiments in which, after the formation of a dominant focus, the active electrode of the direct current is moved to another part of the cortex, to the zone of representation of the hind limb. In that case afferent stimulation frequently continues at first to reinforce the old dominant focus and to evoke contraction of the forelimb muscles, i.e., inertia of the dominant focus is manifested. The stimulation then begins to reinforce the two foci in turn, with predominance of contraction of the muscles alternately in the fore- and hind limb. The current must be switched off for about 30 minutes, and not until then does moving the electrodes into the hind-limb area lead to the formation of a new dominant focus. Before removal of the current and complete extinction of the first focus, the second focus could not be created, because dc polarization of the hind-limb area merely reinforced the existing dominant focus in the forelimb area.

These experiments show that a focus of excitation produced by polarization with a weak direct current in the rabbit's sensorimotor cortex possesses all the properties of a dominant focus. Such a focus can increase its own excitation in response to various forms of stimulation (acoustic, photic), and it can also simultaneously inhibit other cortical regions. A focus of excitation produced in the cortex by a weak direct current also possesses stability and inertia — the ability to maintain excitation once it has begun even when the source of stimulation is removed. In some experiments a dominant focus was observed after stimulation of the sensorimotor cortex by an induction current. Evidently any factor leading to the formation of a stable focus of excitation can give rise to a dominant in the rabbit's cortex. At the same time it must be emphasized that a weak direct current is the most adequate of all artificial stimuli for producing a focus of excitation, since it most closely resembles in character, and the mechanism of its effect, the slow potentials in the central nervous system.

* * *

The experiments cited above show that if a focus of excitation exists in the rabbit's motor cortex, afferent stimulation will evoke contraction of the limb muscles. As has repeatedly been stated, the mechanism of the dominant focus plays a particularly important role in the first stage of conditioning, the phase of generalization (Asratyan, 1937, 1938, 1958, 1963). In the phase of generalization no strictly stabilized CR connections have yet been formed, and any stimulus of the same or another modality can evoke a defensive, food-getting, or other reflex. In other words, the focus of increased excitation responds to impulses regardless of where they arise, i.e., it exhibits the properties of a dominant focus.

The close similarity between the physiological mechanisms of formation of the dominant focus and the mechanisms of formation of temporary connections is emphasized again by the fact that reflex movement of the limbs in the short-term experiment also takes place during the afterperiod of action of the direct current, up to 20-30 min after its removal.

Of all its properties it is inertia in which the dominant most closely resembles temporary connections. In Ukhtomskii's experiments a "rubbing" dominant evoked by local application of strychnine to the frog's spinal cord lasted for 40-60 min, but it could be restored on the day after poisoning (Ukhtomskii, 1950).

The inertia of the cortical dominant produced by a weak direct current in rabbits is long even under short-term experimental conditions.

2. Conversion of the Dominant into Inhibition; the Double Strength Optimum of the Direct Current

Investigations have shown that during weak anodal direct current (1-3 µA) on the surface of the rabbit's motor cortex, in the area of one limb, in response to afferent stimulation (photic or acoustic) to which the orienting reflex had previously been extinguished, a motor response, entirely or mainly affecting the corresponding limb, is obtained (Novikova, Rusinov, et al., 1952; Sokolova, 1954, 1958, 1959; Pavlygina, 1956, 1962, 1964). These findings were confirmed by Morrell (1961, 1962) Hori (1965), and others. This response is easily inhibited. Increasing the strength of the direct current is sufficient to convert the dominant into inhibition.

A more detailed investigation of the conversion of a cortical dominant into inhibition showed that with a gradual increase of the direct current from 0 to 10 µA the response—strength curve is bimodal (Kalinin, 1965). With a gradual increase of the direct current, in response to the same afferent stimuli there is at first no motor response, but when 1-2 µA is reached the response begins to appear. It increases in amplitude as the current increases to 3-4 µA. A further increase in current, however, inhibits the response, which reappears during polarization of between 7 and 10 µA, only to become inhibited again if the strength of polarization goes beyond this point. In the dc polarization experiments there is no strictly definite strength of current, common to all animals, which corresponds to the position of the first and second maximum of the curve. Even for the same animal, the required strength of current varies from one experiment to another.

The fact that there are two optimal strengths of direct current for obtaining a motor reflex suggests that at the site of polarization of the motor cortex there are two structures which play a direct part in the formation of the dominant and which enable impulses to pass from afferent systems (visual, auditory) into the focus of polarization, with its exit to the pyramidal tract.

What are these structures in the motor cortex, which usually do not receive impulses from photic or acoustic stimulation, but which when polarized begin to summate the effects of these stimuli and to give a motor response?

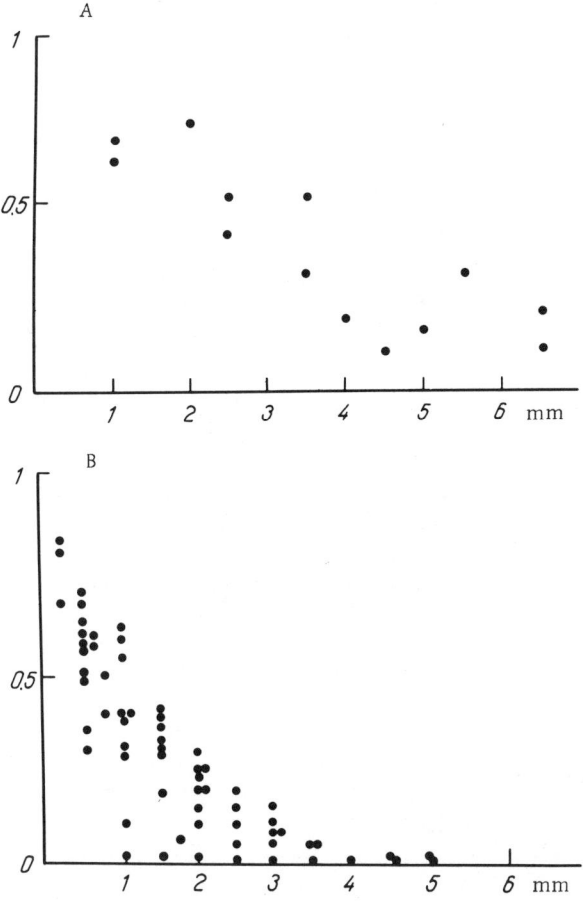

Fig. 14. Distribution of potential during anodal polarization of motor cortex by 2 μA direct current (work of Kuznetsova). A) Distribution of potential over surface of rabbit's cortex; B) in depth of cortex. Potential at point of polarization taken as unity. Each point gives mean value for experiment. Abscissa, distance from test point to polarizing electrode; ordinate, potential at test point.

To answer this question, Kuznetsova investigated the distribution of potential during weak polarization of the motor cortex. The results of her experiments are given in Fig. 14. They show that during polarization of the motor cortex with a weak, direct anodal current strong enough to produce a focus with dominant properties, the potential is reduced by 50% 2-3 mm from the point of polari-

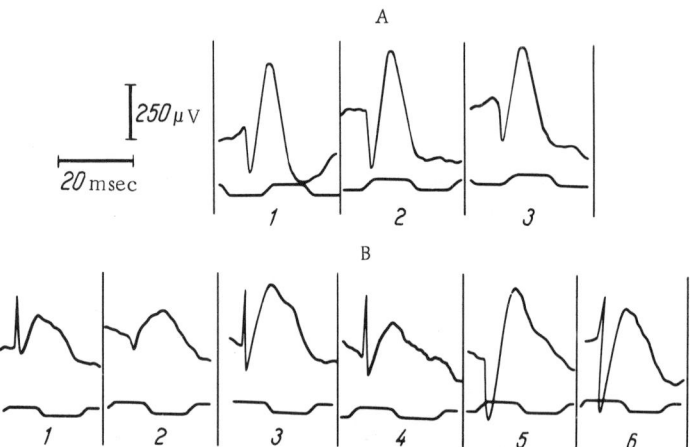

Fig. 15. Dendritic potentials and their changes during polarization of the rabbit's cortex with weak anodal current (work of Kuznetsova). A) Before polarization, in response to stimulation of the cortical surface: 1) 8 V, 2) 8 V, 3) 6 V; B) polarization with direct current (4 μV); dendritic potentials in response to stimulation: 1, 2) 4 V; 3, 4) 6 V; 5, 6) 8 V; square pulses, 0.2 msec, from Neurovar stimulator. Distance between stimulating and recording electrodes 2-3 mm, electrocortical polarization between them, 0.5 mm to one side. Increase in duration of dendritic potentials and accessory potentials on descending phase, with an initial increase, followed by a decrease in their amplitude, are visible against the background of cortical polarization.

zation, and reaches zero only 4-5 mm from it. As Fig. 14 shows, this potential gradient is steeper in the depth of the cortex than on its surface; the potential is reduced by 50% only 0.5-1 mm away from the surface. Consequently, polarization affects predominantly the surface layers (I and II) of the cortex.

Most of the structural elements in layer I of the cortex consist of apical dendrites of pyramidal and fusiform neurons and a system of axonal fibers. The greatest horizontal extent of the apical dendrites, according to Chang (1951), is 4 mm, while according to other authors it is much less. It is interesting to compare the values: the distribution of potential over the cortical surface reaches its minimum (0.1 of the value of the potential at the point of polarization) 4 mm away from the point of polarization, i.e., the greatest length of the apical dendrites. For polarization with a weak direct current, it must be assumed that mainly the apical dendrites are affected. During polarization changes are observed in the shape of the dendritic potential in response to stimulation of the cortical surface: an additional negative wave is obtained, similar to that observed in response to much stronger stimulation, evoking a dendritic potential without application of the direct current (Fig. 15). A number of hypotheses have been put forward to explain the origin of these additional negative waves, usually observed in the descending phase of the dendritic potential. In particular, it has been suggested (Roitbak, 1955; Okudzhava, 1963) that these waves are due to impulses transmitted to apical dendrites through interneurons, which are particularly numerous in layer II.

The two structures playing the principal role in forming the focus of excitation during polarization of the cortical surface with a weak direct current are evidently the apical dendrites of the pyramidal cells and the interneurons.

The reticular system is known to be connected with the pyramidal cells through interneurons, which increase the excitability of the pyramidal neurons under the influence of stimuli of different modalities. It may be asked whether the double optimum of strength of the direct current is a feature peculiar to the motor response obtained during polarization of the motor cortex, or whether the phenomenon reflects some general principle. Is there a double optimum of the strength of the direct current in other responses during polarization of other regions of the cortex?

Gustson (1964) polarized the rabbit's visual cortex with 0.25 to 10 μA and recorded evoked potentials to photic stimulation in the cortex and in the thalamic relay nuclei. The basic form of the primary response of the cortex and lateral geniculate body during long-term experiments was the same as in short-term experiments. Primary responses in the cortex had a latency of between 17 and 22 msec, and in the lateral geniculate body between 15 and 20 msec. Application of anodal current of increasing strength gave

CHAPTER III: A MODEL OF THE CORTICAL DOMINANT FOCUS

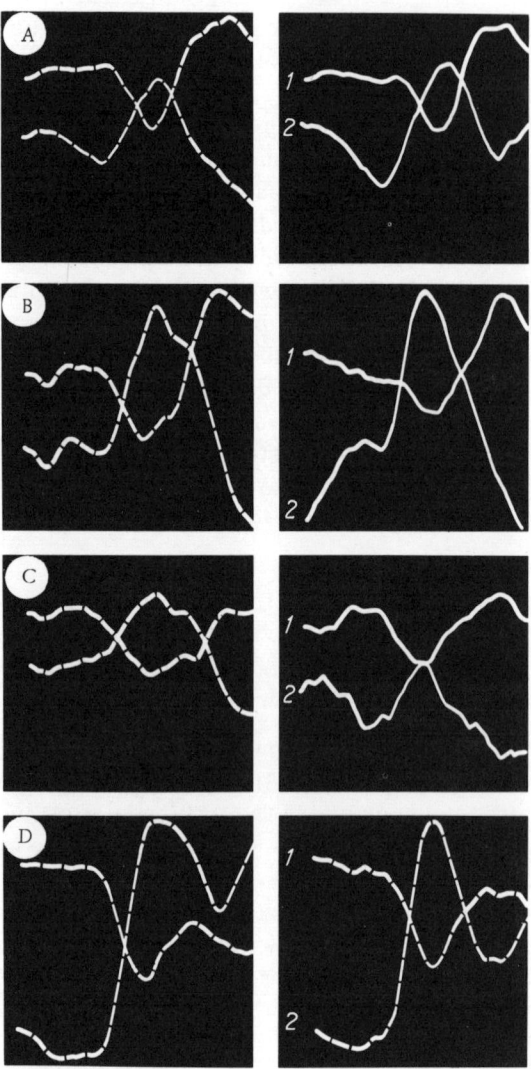

Fig. 16. Changes in evoked potentials in cortex (1) and lateral geniculate body (2) of a rabbit during anodal polarization of visual cortex (work of Gustson). A) Before polarization; B) polarizing current 2 μA; C) 3 μA; D) 5 μA; Calibration: cortex 25 μV, lateral geniculate body 25 μV; beam of CRO interrupted every 5 msec.

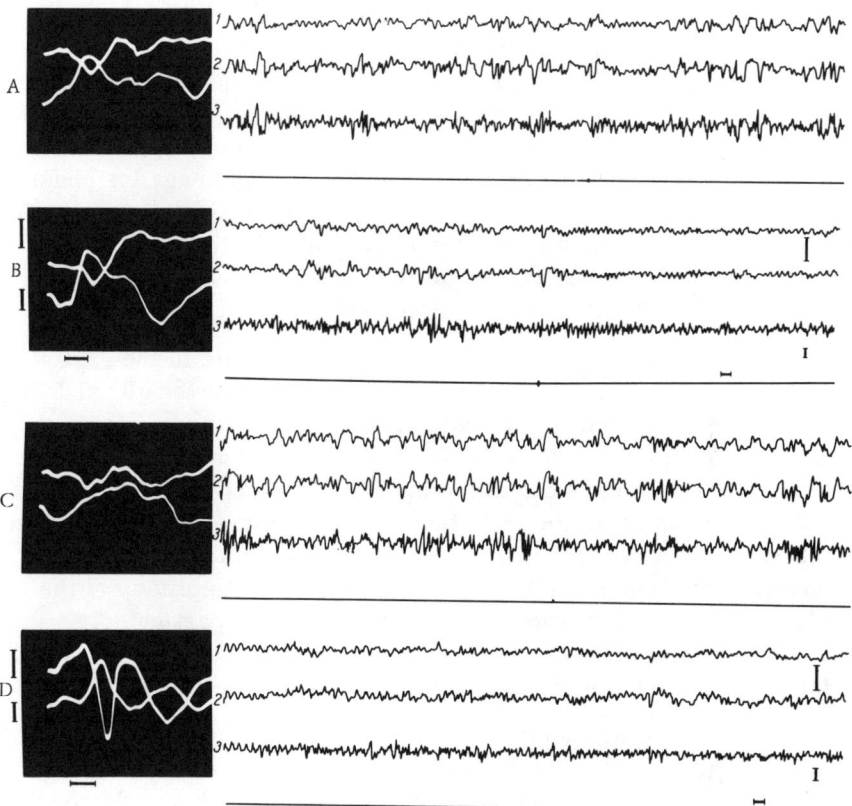

Fig. 17. Changes in evoked potentials of visual cortex (top trace, left) and lateral geniculate body (bottom trace, left) and global EEG (right) during anodal polarization of visual cortex (work of Gustson). A) Before polarization; B) polarizing current 0.25 μA; C) 1 μA; D) 3 μA. 1) EEG (bipolar) of visual cortex; 2) EEG (monopolar) of visual cortex; 3) electrogram of lateral geniculate body (monopolar). Calibration of responses: cortex) 100 μV; lateral geniculate body) 100 μV; time marker 10 msec; calibration of EEG: cortex 200 μV, lateral geniculate body 250 μV.

the following changes in responses in the cortex and lateral geniculate body (Fig. 16). With 2 μA there was an increase in amplitude of the negative phase of the cortical response. The positive phase of the cortical response remained unchanged or decreased. There was a parallel increase in amplitude of the response in the lateral geniculate body (Fig. 16B). With an increase to 3 µA, a decrease in amplitude of all components of the cortical response was recorded. In the lateral geniculate body the amplitude of the evoked response to photic stimulation also was reduced (Fig. 16C). If the current was increased further (to 5 µA), the amplitude of the positive phase of the primary cortical response increased. The negative phase remained depressed. The response in the lateral geniculate body at this time was again increased (Fig. 16D). With a further increase in the current, evoked activity was depressed both in the cortex and in the lateral geniculate body.

Parallel with the changes in evoked potentials in the cortex and thalamus, changes in amplitude and rhythm of the global EEG took place (Fig. 17). No definite EEG pattern could be obtained for the maximum or minimum of the excitation/strength curve, but the transition from one state to the other was clearly revealed on the EEG. For instance, the transition from inhibition to excitation was marked by a shift of the EEG toward fast rhythm. During the change from inhibition to excitation, increased amplitude of the slow waves was observed. These changes were particularly marked during the transition from the first excitation to the first inhibition, and from the first inhibition to the second excitation.

In some experiments involving polarization with a direct current of between 0.25 and 10 µA, not two optima but only one was obtained. In such cases the changes in response due to the anodal current differed from that described above in that it began, not with excitation, but with inhibition as a result of the action of the current. For 10-15 min after the beginning of polarization, instead of an increase in the intensity of these responses in the cortex and lateral geniculate body, there was marked depression of the evoked responses. A further increase in the direct current led to an increase in the positive phase of the cortical response, with a simultaneous increase in amplitude in the lateral geniculate body. The negative phase of the cortical response remained depressed or was absent. If the direct current was increased still further, depression of both the cortical and thalamic responses to photic stimulation was recorded.

The results show distinctly that changes in the shape of the cortical response are accompanied by changes in electrical activity of the lateral geniculate body. Before polarization the cortical

2. CONVERSION OF THE DOMINANT INTO INHIBITION

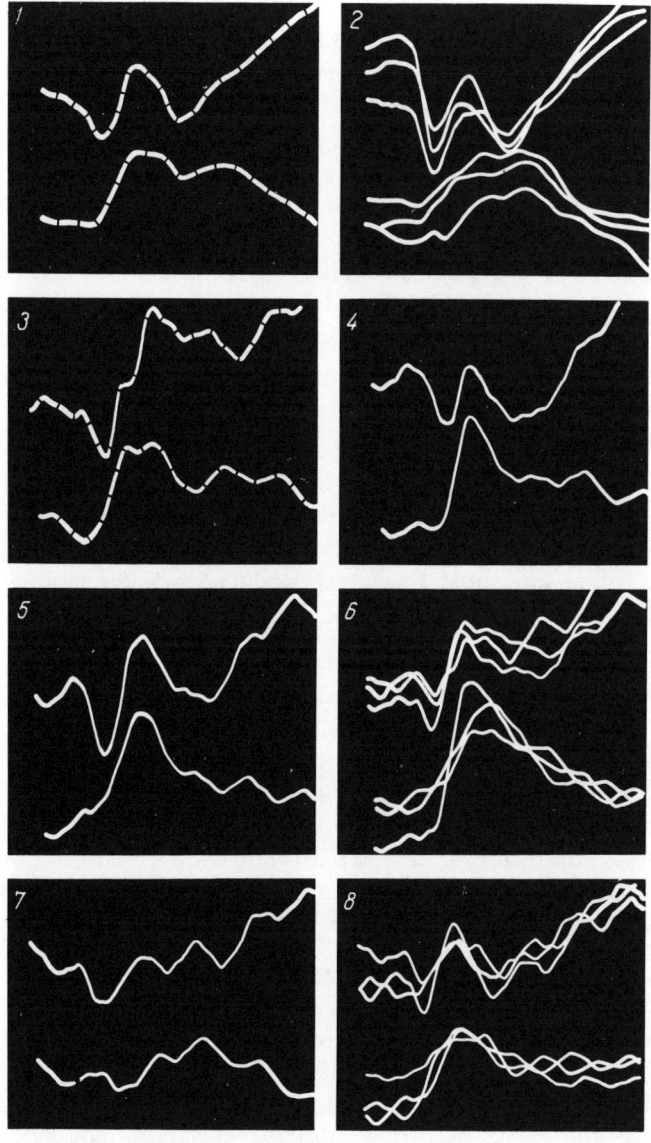

Fig. 18. Effect of changes in cortical response on shape of evoked potential of lateral geniculate body during polarization of visual cortex (work of Gustson). 1, 2) Background; 3-6) polarizing current 0.25 μA; 7, 8) 1.0 μA. Calibration: cortex 50 μV, lateral geniculate body 50 μV, time marker 5 msec.

response to photic stimulation consisted of a triphasic wave (positive—negative—positive) or of two consecutive positive waves the second of which was seen on the rising slope of the first wave (Fig. 18, 2). The response in the lateral geniculate body was unusually shaped and consisted of a prolonged, low-amplitude negative wave. During anodal polarization at 0.25 μA instead of the triphasic response a biphasic, positive—negative potential was recorded in the cortex, i.e., the primary response in the cortex no longer differed from those described in the literature. During this time an increase in amplitude and decrease in duration of the response to photic stimulation were observed in the lateral geniculate body, so that in principle it did not differ from those recorded usually (Fig. 18, 3-5).

No definite strength of current common to all animals could be obtained for the first or second maxima of excitation or minima of inhibition. The first maximum occurred within the range from 0.25 to 3 μA, and the second between 3 and 9-10 μA. The first maximum was most frequently induced by currents of between 0.25 and 1 μA, and the second by currents of between 3 and 5 μA.

The first maximum of excitation was characterized by a statistically significant increase in amplitude of the negative phase of the cortical response ($t = 3.875$; $P < 0.001$). The positive phase of the response showed no statistically significant changes. At this time there was a significant increase in amplitude of the response of the lateral geniculate body ($t = 4.2$; $P < 0.001$). During the phase of inhibition, there was a significant decrease of all components of the cortical response and of the negative phase of the response in the lateral geniculate body.

During the second phase of excitation a significant recovery of amplitude of the positive phase of the cortical response to its background level is observed (difference from the background level not significant), while the amplitude of the negative phase remains low, just as during the period of inhibition. At this time the increase in the response of the lateral geniculate body by comparison with the background is statistically significant.

In the experiments in which direct current on the cortex produced a different set of changes, starting with depression, the degree of inhibition at the onset of the current also was statistically significant: a decrease of the cortical and thalamic responses was observed. During the period of excitation, there was a significant increase in the positive phase of the cortical response (with no change in the negative phase). The response of the lateral geniculate body was significantly increased over the background level.

Experiments in which the rabbit cortex was anodally polarized demonstrate the role of the visual cortex in regulating conduction of specific afferent impulses through the lateral geniculate body. Whatever the functional state of the cortex induced by weak polarization, definite changes take place in activity in the lateral geniculate body: stimulation (or depression) of evoked potentials in the cortex correspond to stimulation (or depression) of responses of the thalamic relay nucleus.

These results are related to the problem of corticofugal connections with various subcortical structures at present undergoing extensive study in Soviet and other laboratories. Some aspects of this problem have been examined in detail by Narikashvili (1962) and Meshcherskii (1966), and, in particular, changes due to the action of direct current have been investigated by Gustson (1964).

Another problem has also been examined: that of the double strength optimum of the direct current when acting on an animal's cerebral cortex. Let us consider the results of these experiments in the light of this other problem.

The results obtained by polarization of the visual cortex are in agreement with those obtained by polarization of the motor cortex. With a gradual increase in current and application of the same afferent stimulus, in both cases there are two optima. Since the double optimum of strength of the direct current within the range 0.25-10 μA occurs not only during polarization of the motor cortex in motor responses to afferent stimulation, but also during polarization of the visual cortex in evoked responses, the double optimum is apparently a general rule for the action of a direct current on the cortex.

Experiments in which no double optimum was obtained in the visual cortex do not contravene the general rule, because in these experiments, in which there was only one optimum, the onset of the direct current (starting from 0.25 μA) is always marked by a phase of depression of cortical and thalamic responses. This may imply that the first optimum of current strength in these rabbits was below 0.25 µA.

What is the explanation of the two optima during anodal polarization of the cortical surface? One possible explanation is that in the cortex there are two different structures which respond at different times by a change in their state (Rusinov, 1961). Kuznetsova studied the distribution of potential during weak polarization of the rabbit's motor cortex and the change in the shape of the dendritic potential in response to increases in the polarizing current. On the basis of her results it was postulated that the two structures playing the principal roles in the formation of a focus of excitation in the area are the apical dendrites of the pyramidal cells and interneurons, and that the activity of each has its own characteristic optimal level of direct current (Rusinov, 1965a, b).

It must also be borne in mind that the double optimum during dc polarization is also found in the peripheral nervous system, if an altered focus in the frog's sciatic nerve is anodized during recovery of its conducting function (Rusinov, 1934). Consequently, the double optimum during the action of weak direct current is a characteristic phenomenon of the whole nervous system, both central and peripheral.

The cortex consists mainly of cells of two types: neurons and glia. From the point of view of energetics, neurons and the perineuronal glia are known to constitute a single functional system (Kuffler et al., 1966). When functional activity is increased, the neuroglial cells move toward the neuron and the number of glial cells surrounding the neurons increases (Aleksandrovskaya et al., 1965). The view originally expressed by Golgi, that glial cells form an intimate link between capillaries and neurons, is widely held. Electron-microscopic findings confirm the hypothesis of the perineuronal glia as an intermediate link in the path taken by metabolic products from blood vessels to neurons. Although Kuffler (Kuffler et al., 1965) concluded that substances are transported from the blood stream to nerve cells, not by glial cells but along the interstitial spaces, nevertheless the possibility is not ruled out that bodies of more than a certain diameter, and thus unable to diffuse through the narrow interstitial spaces, may be ingested by glial cells and carried to neurons. The precise measurements of the intercellular spaces are not yet known, and they may be greater or smaller in the living brain than is observed in the electron microscope.

It is natural to suggest that the two structures in the cerebral cortex which respond at different times to the action of a weak direct current are neurons and neuroglia. If apical dendrites of neurons are mainly concerned in the organization of the first optimum, it is the glia which is mainly concerned with the organization of the second optimum during mobilization of the interneurons into the response. In that case, this point of view would not be contradicted by the existence of a second optimum of strength of the direct current in the peripheral nervous system during restoration conduction in an altered focus of a nerve on anodization. The two structures responding at different times to the weak direct current in this case must be the axons and glial cells, remembering that the Schwann cells constitute the peripheral neuroglia.

Polarization of the nervous system by a weak direct current can be used as one method of obtaining a physiological model of successive recruiting of the structures of the nervous system into the response to stimulation.

The events taking place in the motor cortex during conversion of a focus of excitation into a dominant characterize only one aspect of the system formed during the reaction. This system also includes the specific cortical area of representation of the afferent stimuli reinforcing the focus. Events taking place in this particular cortical area can be judged, at least in part, from evoked potentials in response to afferent stimulation. Examining what happens to the evoked potentials in the visual cortex when a focus of excitation in the motor area responds by a reflex to hitherto indifferent visual repetitive stimulation, it can be asked whether the visual cortex itself plays an active role in the motor response of the dominant focus to photic stimulation.

It will be clear from Fig. 19 that during polarization of the motor cortex, an evoked potential with a large late negative phase is recorded in the occipital region in response to each flash. This picture is observed before the reflex motor response. Just before the beginning of the motor response, however, the evoked response starts to change its shape: its large negative and, evidently, sec-

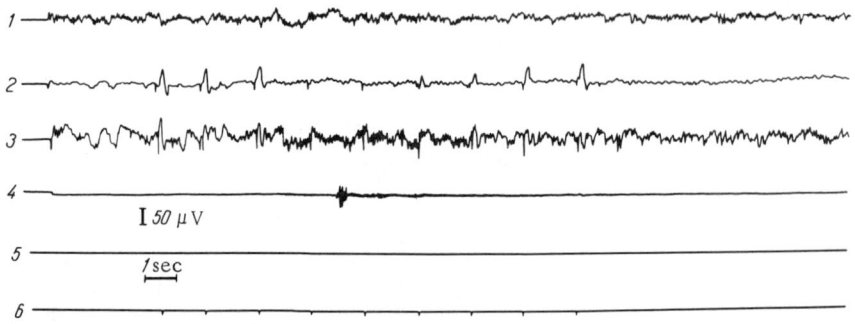

Fig. 19. Inhibition of evoked potentials to flashes in the rabbit visual cortex when a dominant focus is present in the motor cortex on the right side (work of Kalinin): 1) EEG of right motor cortex; 2) flash-evoked potentials in right visual cortex; 3) electrical activity of lateral geniculate body; 4) movement of left forelimb; 5) movement of right forelimb; 6) marker of stimulation.

ondary wave begins to be inhibited; its initial positive response still remains. The secondary negative waves of the evoked responses to photic stimulation in the occipital region are not restored immediately after the end of the reflex movement, but gradually.

These facts show that during the motor response of a dominant focus to photic impulses, secondary evoked potentials in the occipital region are inhibited. In the presence of the dominant focus, the visual cortex is affected by events in the motor area. The dominant focus in the motor area is reinforced not only by impulses of different modalities traveling along the path of the reticular system directly or via interneurons to the pyramidal cells, but the specific cortical area of representation of the reinforcing stimuli also plays an active role.

* * *

The model of a dominant focus formed by polarization with a weak direct current can be used to investigate the excitable and conducting structures of the central nervous system which are successively recruited into the response of the focus of excitation. The same motor response of a dominant focus is effected by two different structures in turn during polarization by a current of between 0 and 10 μA, so that the reliability of its occurrence is ensured.

3. Trace Phenomena Associated with the Dominant Focus and Memory

Mathematics is being applied increasingly in all fields of science, inclusing the physiology of the nervous system. In particular, the theory of automatic control, based on strict mathematical methods and solving the fundamental problems of analysis and synthesis of modern automation systems, is applicable to the physiology of the brain and of its higher levels as the controlling part of the system. In theory it is assumed that any process of automatic control, irrespective of the complexity of the structure of the system in which it takes place, is characterized not only by a direct link between the controlling and controlled parts, but also by a retrograde link, or feedback, between them.

The mathematician von Neumann (1955), working in the field of automation theory, defines the feedback as the "line responsible for the cyclic path of excitation." A feedback, in the form of a circular rhythm, has long been familiar in physiology. The literature on circus rhythm in physiological structures until 1930 was summarized by Samoilov (1930), and until 1956 by Smirnov (1956). The properties of feedbacks are directly related to the function of foci of excitation in the central nervous system. When analyzing the possible mechanisms of formation of the dominant focus, Ukhtomskii considered the possibility of a stationary cycle, joining together a series of centers with continually renewed waves of excitation, or repetitive impulses. Stationary excitation in a constellation of centers can exist as a cycle in which each segment of this constellation, connected with its neighbors and stimulating them, in turn receives stimulation from them, so that, as Ukhtomskii wrote in 1937: "The cyclic machine thus formed remains for a long time capable of bursting into activity and increasing its momentum under the influence of extraneous and accidental stimuli."

The simplest case of cyclic connection, consisting of a ring formed by an axon collateral with return to its own neuron, according to automation theory is a "memory." If such an organ receives a single excitation at its input, it remains excited and must give out impulses at its output continuously.

The ring of the jellyfish in the experiments of Meyer (1906) and Vetokhin (1929) provides a physiological model of such a memory. Vetokhin's experiment is particularly interesting. The excised ring of the jellyfish *Cyanea arctica,* on which he conducted his experiments, reaches a diameter of 10-20 cm. By an ingenious incision, Vetokhin trebled this diameter. By stimulating the ring at any point, and blocking the spread of the wave of excitation in one direction, Vetokhin was able to observe the wave of excitation traveling in the other direction around the ring at a speed of 15 cm/sec for 5.5 h. This experiment, demonstrating a physiological model of circus rhythm in its simplest form, as the unidirectional movement of an impulse along a substrate without delay at synapses or other relays, is important for the investigation of circus rhythm in the nervous system as a physiological mechanism of memory.

A problem discussed in the literature is whether memory depends on continuous activity of impulses and, in particular, of movement around a ring of neurons, or whether memory is based on structural changes left behind by preceding stimuli.

The experiments of Gerard (1953) with hamsters are interesting in this respect. The animal ran for hours in a maze, after which it was either cooled to a certain temperature, or given an electric shock. In both cases the EEG showed disappearance of background activity. If the shock was produced after running for 4 h, this did not affect the stability of the "training." If the shock was given after running for 1 h or less, this was reflected in the results of training, to an amount which varied inversely with the length of the running period.

Assuming that waves of the EEG reflect a dynamic process in the brain, Gerard regards the mechanism of memory as basically a static process of structural change. He admits an initial circulation of impulses around a ring of neurons only to reinforce the structural changes.

A similar viewpoint is reflected in papers on automation theory. Kleene (1956), examining events in "nerve nets" and in automata on the basis of the nerve models of McCulloch and Pitts (1956), considers that only short-term memory can be explained by the circulation of nerve impulses. In his opinion, the prolonged circulation of impulses must inevitably cause fatigue, and such a mechanism cannot satisfactorily explain long-term memory. Kleene ignores the inhibitory response which, during prolonged circulation of impulses, can develop before fatigue.

From my point of view, foci of excitation in the central nervous system, in the cerebral cortex, are directly related to physiological mechanisms of memory. Activity in foci of excitation in the central nervous system is bimodal and is reflected electrographically as a sustained shift of steady potential and its fluctuations of different form: periodic, as a rhythm, almost periodic, and stochastic. Without dwelling on the genesis of these forms of fluctuation of potential, suffice it to say that in their degree of complexity the foci themselves can be divided into: a) simple foci of excitation; b) foci becoming dominant and bursting into activity in response to any stimulus; and c) foci forming complex systems of temporary connections responding only to adequate stimuli.

The investigation of foci of excitation has revealed factors which show that to fix the rhythm of stimuli applied to an animal experimentally in its central nervous system, a certain length of time is required, but this fact is not new. It was pointed out by Gerard and others. What is new is that this period must be connected all the time with the stimuli. In her experiments on rabbits, Grechushnikova produced a dominant focus by gradually diminishing repetitive electrodermal stimulation of one forelimb. When this stimulation was stopped, she found just as Ukhtomskii and Vinogradov (1925) had found in spinal frogs, that in response to extraneous stimulation stronger contractions of the limb were obtained for the center in which the focus of excitation had been formed, and the contractions were synchronized with the diminishing stimulation. In some cases application of an afferent stimulus 2-3 min after the end of stimulation evoked no motor effect, but the same afferent stimulus, 15-20 min later, with no change in the experimental conditions, evoked a response of the limb corresponding to the dominant focus in a rhythm created by that focus. The decisive factor in this case is time.

The residual contracture after individual contractions indicates that in this case, just as in the experiments on the swallowing dominant or with polarization of the cortex, a focus of stationary excitation, reflected at the periphery as a contracture of the skeletal muscles, is formed.

Why did the contractions in the imposed rhythm not arise a few minutes after the diminishing electrodermal stimulation, but only after a comparatively long time interval? This could be attributed to fluctuations of excitability in the focus, but the excitability itself cannot circulate around the ring, if movement around that ring is associated with the rhythm of excitation. Excitability as a threshold response, even though a very important physiological parameter, itself depends on the functional state of the focus of excitation and is a derivative of that state. This fact likewise cannot be explained purely by the summation of excitation from afferent stimuli and the stationary excitation of the focus, when in response to a series of afferent stimuli a reflex is obtained only to one stimulus. Such a series of stimuli was not used in this case.

This fact must be understood as most likely the result of "finishing" of the rhythm in the central nervous system itself, the finishing implying that waves of excitation from the stimuli in the imposed rhythm, forming a dynamic stereotype in the central nervous system composed of cycles of excitation circulating in succession, are more and more concentrated in one circle of neurons, inhibiting all the rest. In this way, a focus of excitation is formed which is temporarily dominant in the central nervous system and possesses increased excitability and other properties, as a result of which it responds to any continuous afferent stimulus by the previously imposed rhythm of excitation.

It may be asked how long the excitation can continue to circulate or, in other words, whether a system of continuous circus rhythm can be formed? This question was answered by Samoilov (1930) 40 years ago in his paper entitled "The Circus Rhythm of Excitation." He considered several examples showing the presence of such systems of circus rhythm and calculated the approximate path of the excitation in the various excitable structures forming the ring. Samoilov at the same time postulated that the region of the circus rhythm is infinitely wide. "There is no doubt," he wrote, "that the overwhelming majority of reflexes in our body follow a closed pathway." Instead of the concept of the "open reflex pathway" he introduced the concept of "closed reflex" and gave a diagram of a reflex arc in the form of a "ring of excitation."

Samoilov's ideas are very attractive at the present time in connection with the question of feedback, its functional role in the activity of the living organism, the open or ring scheme of the reflex arc, and long traces in the nervous system as the physiological basis of memory.

4. Foci of Excitation in the Central Nervous System Evoked by a Pulsating Current. Pavlygina's Experiments

There is as yet no single theory to explain the mechanism of memory. However, a characteristic feature of existing theories is the acceptance that memory, in its simplest form, is based on trace activity. The dominant focus, one of the principal properties of which is inertia (the ability to preserve an active state in the afterperiod of stimulation) can serve as a model to investigate the nature of trace processes.

Pavlygina (1956, 1960) investigated trace phenomena in the presence of a dominant focus and studied the role of a shift of steady potential in retention. In her investigations the focus of excitation in the motor cortex, unlike other studies with polarization of the cortex, was produced by a weak direct current having, besides its steady component, a sinusoidal component the period of which could be varied (a pulsating current). She used this type of direct current in an attempt to bring the conditions of fluctuation of electrical potential observed in dominant foci closer to natural conditions. Polarization of the rabbit's motor cortex by the pulsating current itself did not evoke motor responses. Limb movements appeared only in response to afferent stimuli (photic, acoustic, tactile) given during polarization. These movements were synchronized with the pulsations of current. This synchronization with the rhythm of pulsation did not take place at once but developed gradually after a series of afferent stimuli against the background of polarization by the pulsating current. Often the reflex movements were double: two movements of equal or different magnitude corresponded to each pulsation of current. They could take place either as the current was increased or decreased during the pulsation. Changing the pulsating current from one frequency to another led to gradual readjustment of the rhythm of driving: at first the movements followed each other in the old rhythm, but later they began to adopt the rhythm corresponding to the new frequency of pulsation.

These phenomena are easily inhibited. It is only necessary to increase the strength of the direct current acting on the cortex to make them disappear, passing gradually through all the stages which occurred during their appearance at the beginning of the experiment, but in the opposite order. Evidence of inhibition is given by the "rebound" response, consisting of increased reflex movements in response to the same afferent stimuli after removal of the current. The longer inhibition and the greater the number of test afferent stimuli during inhibition, the stronger the movements during the "rebound." The following fact is noteworthy: in the period of the "rebound" after inhibition the rhythm of contractions corresponds to the frequency of the pulsating current which acted previously. In other words, during inhibition in the central nervous system, in a dominant focus, the rhythm imposed by the experimenter continues to be stored in a latent state, and it may become manifested after inhibition. This fact is undoubtedly relevant to the mechanisms of memory and formation of the temporary connection.

In most of these experiments the afferent stimuli (acoustic, photic, tactile) were continuous. Application of interrupted afferent stimuli against the background of an established dominant focus upset the previously driven rhythm of limb contractions.

Conversion of the dominant into a reinforced connection, or CR, is an interesting and important problem. In long-term experiments the dominant focus created by a weak pulsating current in the motor cortex is highly stable and persists for several days. In other words, after one day of the experiment the reflex movements synchronized with the pulsating current described above can be obtained on application of afferent stimuli without the need for further application of the pulsating current on the cortex. If a constant interval (1 min) was maintained between applications of the afferent

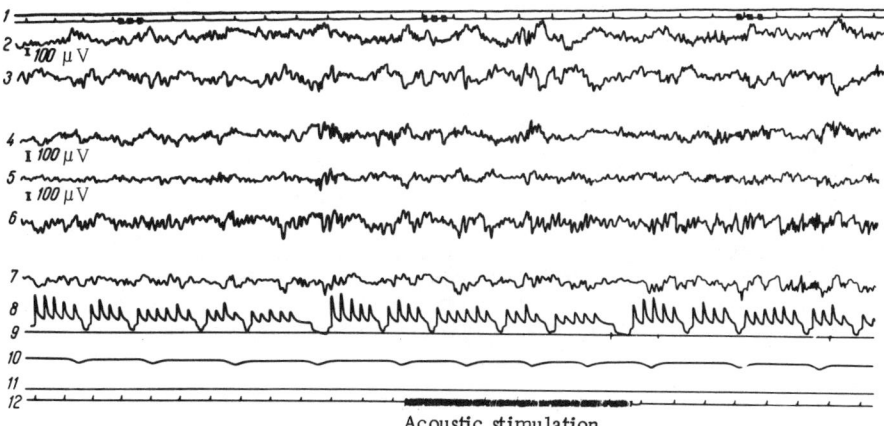

Fig. 20. Cortical and subcortical electrical activity of a rabbit during the action of a pulsating current. Frequency analysis of EEG of motor cortex: 1) time (in sec) and marker of channel for analysis; 2) right visual area; 3) right auditory area; 4) right motor area ("monopolar" recording); 5) right motor area ("bipolar" recording); 6) mesencephalic reticular formation; 7) medial thalamic nuclei; 8) record of analyzer; 9) movements of left limb; 10) respiration; 11) movements of right limb; 12) time (in sec) and acoustic stimulation.

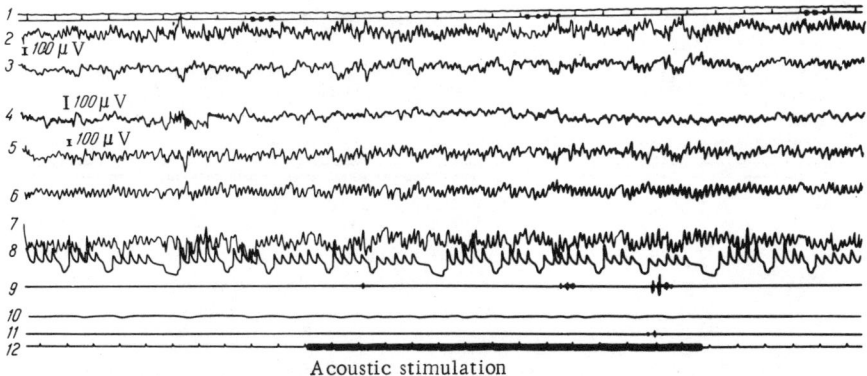

Fig. 21. Appearance of dominant rhythm of 3 Hz in EEG of motor area and also of limb movements in the same rhythm in response to continuous acoustic stimulation during polarization of the right motor cortex by a pulsating current (constant components 0.1 µA, variable 0.2 µA, frequency 3 Hz). Legend as in Fig. 20.

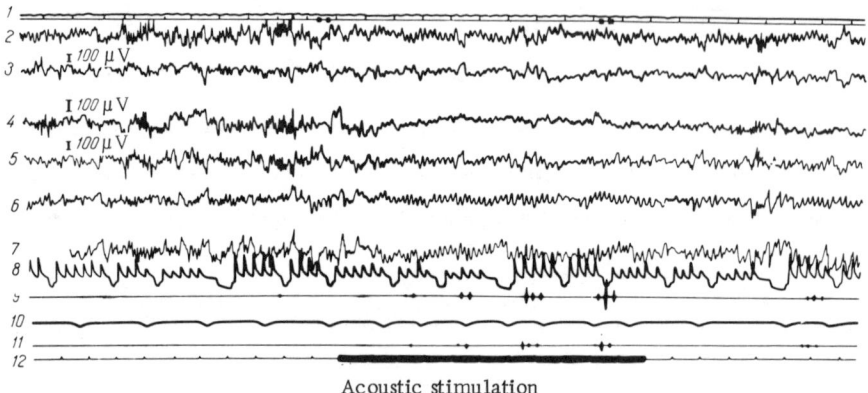

Fig. 22. Analysis of EEG of visual cortex during action of the same pulsating current in the right motor cortex: 1) time (in sec) and marker of channel for analysis; 2) left motor cortex; 3) right visual area; 4) right motor area ("bipolar" recording); 6-12) as in Fig. 20.

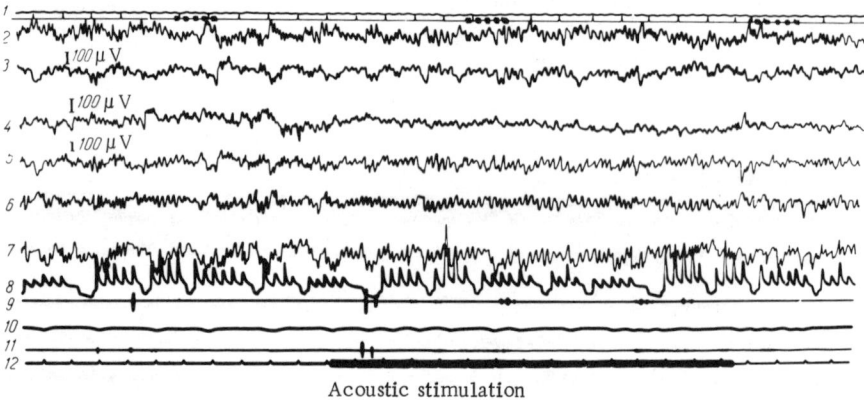

Fig. 23. Analysis of EEG of mesencephalic reticular formation of a rabbit during action of the same pulsating current in the right motor area. Legend as in Fig. 22.

stimuli, a CR to time was formed both during the pulsating current and after its removal. The CR to time was manifested by the fact that rhythmic movements preceded the stimulus or, if one of a series of afferent stimuli was omitted, the movements appeared at that time.

The fact that any response in the living organism can be evoked by a CR is well known. In the case under discussion, however, the manifestation of a CR in the rhythm artificially imposed by the experimenter in the cerebral cortex must be emphasized.

In what structures of the central nervous system does this rhythmic driving by the pulsating current take place: in the cortex itself, or in the reticular system? The answer to this question is given by investigations of the electrical activity of the cortex, the mesencephalic reticular formation, and the medial nuclei of the thalamus. The results show that the rhythm of the pulsating current is followed by the cortex but not by the mesencephalic reticular formation. Frequency analysis (Fig. 20) shows, before and during the afferent stimulus, that no single rhythm predominates in the EEG of the motor cortex. The EEG of the same rabbit, recorded while applying a pulsating direct current to the motor cortex, is illustrated in Fig. 21. The response to acoustic stimulation is shown on the curve. The results indicate that rhythmic driving by the pulsating direct current takes place in the motor cortex whether the recording is "monopolar" or "bipolar." Excitation by acoustic or photic stimuli reaches the threshold level and produces an effect. A motor response appears in the rhythm of 3 contractions per second, i.e., synchronized with the pulsating current. Analysis of the EEG of the visual cortex and mesencephalic reticular formation recorded under the same conditions shows that these structures do not follow the rhythm of the pulsating current (Figs. 22 and 23).

After discontinuation of the current, the imposed rhythm of 3/sec persists for a long time, as is clear from the EEG of the motor area and reflex movements in response to afferent stimulation. If the motor dominant for some reason or other was not formed, the rhythm disappeared immediately after the current was stopped. Rhythmic driving in the cortex usually preceded the movements to afferent stimulation. The manifestation of the dominant focus after removal of the pulsating current is illustrated in Fig. 24.

In all investigations of the action of a weak direct current on the cortex or subcortical structures, a noteworthy feature is the occurrence of excitation and depression depending on the strength

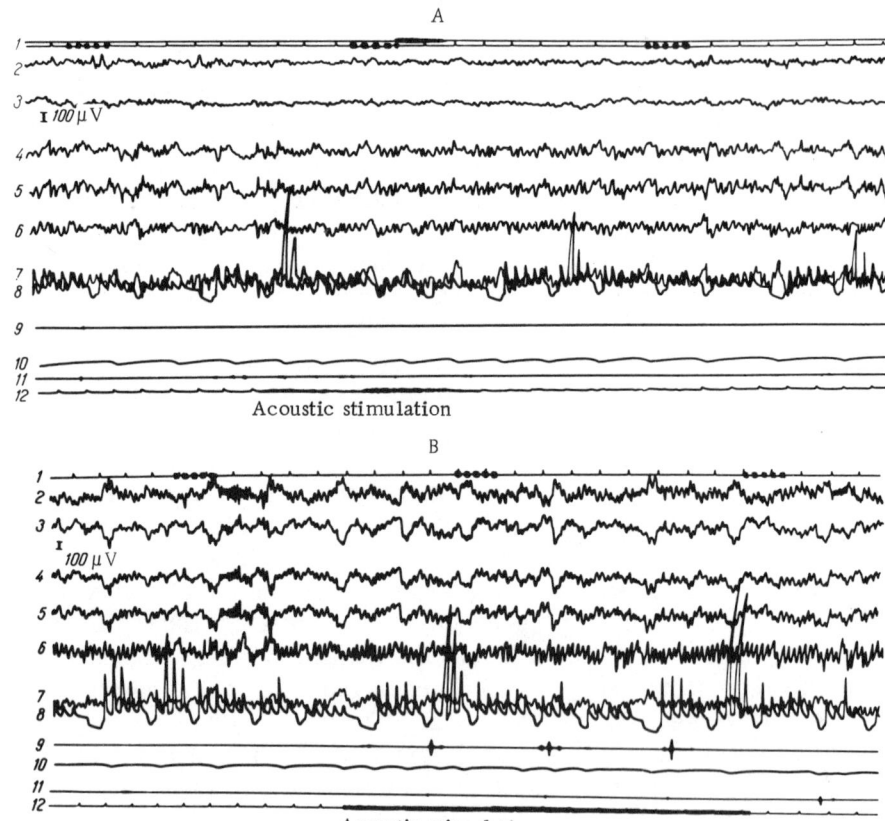

Fig. 24. Maintenance of the driven rhythm in a dominant focus: A) the focus is inhibited by an increase in current. B) current removed. Reflex movements appear to afferent stimulation (continuous sound) in a rhythm of 3/sec, corresponding to the rhythm of the pulsating current before inhibiting. 1) time marker and marker of channel for analysis; 2) auditory cortex ("monopolar" recording); 3) visual area ("bipolar" recording); 4) motor area, symmetrical point ("monopolar" recording); 5) motor area ("monopolar" recording); 6) reticular formation ("bipolar" recording); 7) thalamus; 8) record of analyzers; 9) left forelimb; 10) respiration; 11) right forelimb; 12) time (in sec) and marker of stimulation.

of the current. With a very slight increase in current above the optimum level, the phenomena of the dominant focus described above are not strengthened but inhibited.

4. FOCI EVOKED BY A PULSATING CURRENT

In the investigations with pulsating current, in which the constant and variable components could be varied independently, a rhythmic current without a constant component was less effective than the same current of the same frequency and amplitude of pulsations, but with a small constant component. On the other hand, to obtain the same effect, a stronger direct current was required if it had no variable component. The optimal strength for creation of the dominant focus was a current with a constant component of 0.2-0.3 μA and variable components of 0.1-0.2 μA.

These results indicate the independent roles of the constant component and the rhythmic pulsations at the cortex or subcortical structures and they determine the parameters of current converting the dominant focus from its optimal state into an inhibitory state. If the strength of the smallest direct current (without variable component) required to inhibit the dominant is compared with the constant component of the pulsating current inducing inhibition, in the first case a current of 3-5 μA is required, but in the second case only 0.4-0.5 μA.

The investigation of trace processes shows how accurately and how long the nervous system can reproduce processes taking place previously under the influence of agents acting directly and, in particular, they show how accurately phenomena characteristic of the dominant focus can be reproduced in trace processes. The experiments show that summation persists for some time after the current is stopped. The focus of excitation retains its rhythmic nature. Records obtained after removing the current show that in response to an acoustic stimulus the dominant frequency of the motor area is that which acted previously (3 Hz), and contractions of the limb begin to take place in the same rhythm.

Characteristically not only the imposed form of excitation, with definite frequency characteristics, but also the place of its appearance remain the same. The corresponding rhythm was recorded in the electrical activity of the motor cortex. In Pavlygina's experiments it was displaced into the nonspecific structures of the subcortex, to which a special role in the manifestation of memory has been ascribed.

Analysis of the frequency spectrum of the electrical potentials of the mesencephalic reticular formation some minutes after removal of the pulsating current showed that in response to acoustic stimulation rhythmic movements of the corresponding limb appear, i.e., trace phenomena are present but rhythmic driving is absent in the reticular formation.

It can be concluded that a polarized system of nerve cells can follow the parameters of a stimulus precisely and can thereby retain a characteristic model of the stimulus. At the time of reinforcement of the focus, predominance of the frequency which corresponds to movement of the limb becomes predominant in the electrical activity of the motor area, while slow waves of approximately the same duration as the period of stimulation appear in the cortex.

These results, showing that summation can take place in the cortex if the time interval between stimuli is of considerable duration, are evidence that a form of excitation permitting long trace phenomena exists in the central nervous system. These are most likely to be slow waves of electrotonic character.

Characteristically, if trace processes are to take place not only is the rhythmic nature of the focus of excitation preserved, but also its definite electrotonic gradient. The appearance of trace processes is connected with a certain level of steady potential: slow waves in the EEG and movements appear in response to acoustic and photic stimuli when the level of the steady potential in the motor cortex is between +2 and +4 mV.

If optimal conditions were chosen for manifestation of the dominant focus, the trace phenomena were stable and could not be abolished at a certain stage of their development by the direct action of another stimulus on the structures in which the traces were exhibited. With a change in frequency of the polarizing current the potentials at the point of stimulation retained their previous frequency for a time, despite the action of the other stimulus, and contractions of the limb followed in the same rhythm. Later a new frequency appeared in the EEG, but the movements followed in the previous rhythm, and did not change to the new rhythm until a short time had elapsed.

An interesting experiment was carried out on a rabbit in which a dominant focus was formed by a current of 3 Hz, after which a current of 1.5 Hz was applied. At the moment of reinforcement of the focus the new frequency appeared in the EEG, but the limb continued to contract at the previous rhythm. This dissociation between the frequency of the driven rhythm in the cortex and movement is characteristic of the first stage of transformation of the dominant, and it evidently indicates that in this particular case the deeper neuronal structures retain the old rhythm (and impulses reach the periphery in this form) longer than structures directly influenced at that moment, and from which electrical activity is recorded.

The stability of trace processes was revealed by another type of experiment. With an increase in current the dominant was converted into an inhibitory state: despite the application of additional stimuli, no limb movements were observed. After removal of this current, as was pointed out above, phenomena characteristic of the presence of the dominant focus were restored. The longer the inhibition and the more test stimuli were applied during inhibition, the stronger were the after-phenomena during the "rebound."

During inhibition the outflow to the periphery is blocked and the imposed rhythm continues to remain in the dominant focus. The "learning" process is perhaps not merely preserved, but may actually continue against the inhibitory background, because the intensity of the phenomena associated with the dominant focus was to some extent dependent, after the strong current had been removed, on whether or not test stimuli had been applied against the background of the inhibitory state.

The duration of the trace phenomena depended on the previous stability of the dominant focus. If a dominant was formed repeatedly in the same animal, the trace phenomena could be maintained for a long time and reproduced even on the day after the experiment. During fixation of traces, the whole situation surrounding the animal was evidently included, for trace phenomena in the form of a definite rhythm in the potentials of the motor cortex could arise as a CR simply from placing the animal in the chamber.

Longer preservation of trace processes was facilitated by the stereotyped character of the experiment. In some experiments the stimuli reinforcing the dominant focus were applied at fixed time intervals (1 or 2 min). Under these conditions summation phenomena were highly regular. If photic and acoustic stimuli continued to be applied after removal of the current at the same time interval as while the current was being applied, the trace phenomena lasted much longer than if these stimuli were applied irregularly.

Introduction of an element of temporary memorizing thus did not interfere with the manifestation of the trace processes but, on the contrary, facilitated their fixation.

After removal of the current, the trace processes continued to appear for some time and then were extinguished. If the manifestation of this process is analyzed relative to individual elements, a number of stages can be distinguished which very greatly resemble the stages observed during formation of the dominant, except that they are in the opposite order.

During formation of the dominant focus, in the first minutes after onset of the pulsating current no significant changes were observed in the EEG. Only after a number of successive applications of the stimuli reinforcing the focus did potentials in the rhythm of the pulsating current appear in the motor area to the photic or acoustic stimulus. The movements at this time could be single, and after a while they began to resemble the imposed rhythm, but most frequently movements in a slower rhythm were observed. Later, contractions of the limb in the rhythm of the applied stimulus appeared. During extinction of the trace processes, the peripheral part of the motor system was inactivated first: the movements lost their precision and as a rule they were slower and appeared less regularly. Trace phenomena in the form of following the corresponding rhythm in the EEG of the motor area, on the other hand, persisted much longer.

The apparently gradual adaptation of the motor system to the imposed rhythm observed during formation of the dominant focus is thus evidence of differences in lability of the central and peripheral parts of the motor system. It is difficult to say at present where the delay takes place: whether it is due to slower assimilation of the rhythm by the pyramidal cells or to neuromuscular transmission.

In Pavlygina's experiments the cortical surface was stimulated and the potential recorded from it. Recording the EEG showed that cortical rhythmic driving takes place much sooner than contractions of the limb appear.

4. FOCI EVOKED BY PULSATING CURRENT

Traces in the nervous system from previous stimuli are the physiological basis of memory. Trace processes are a common property of the nervous system and there is no reason to suppose that special centers of memory exist. Traces in the central nervous system may differ in duration and complexity, and this is reflected also in the different classifications of memory.

The dominant focus, like any system with feedback, can be regarded as one mechanism of memory, as is clear from experiments with polarization of the animal's cerebral cortex by a pulsating direct current. As was stated above, the best condition for creation of the dominant focus was application of a current with a constant component. Trace processes were totally dependent on the dominant focus in their manifestation: the stronger the dominant focus, the longer and more regularly the trace processes appeared. Hence it can be concluded that polarization of nerve cells or, more precisely, a change in the level of their polarization to the optimum, facilitates trace retention. This view is confirmed by data in the literature. Morrell (1962a, b) polarized the visual cortex by a direct current. With a microelectrode he recorded the activity of a cell which responded to 3/sec flashes. Under these conditions the cell also responded to a single flash applied 20 min after the repetitive stimulus by a series of discharges also at 3/sec. Grouping of the discharges of a single unit during interrupted stimulation was also observed by Jasper et al. (1962) and by Strumwasser and Rosenthal (1960). However, these authors do not find the rhythm of stimulation maintained in the absence of the stimulus. This is evidently the result of polarization of the cell.

Trace activity in the presence of a motor dominant focus was characterized by preservation of a definite model of the stimulus: preservation of the frequency characteristics with the corresponding change in level of the steady potential. When Morrell (1962a, b) repeated the cortical polarization experiments, he found that "in this phenomenon just as in learning," memory can also be preserved in a particular sensory system. For instance, if only acoustic stimuli, and never photic stimuli, were applied during the subthreshold current in the motor area, after removal of the current the focus was reinforced only by acoustic stimuli. These experiments, like the earlier findings of Sokolova (1959), show that a system of polarized cells retains for a long time not only the characteristic model of a given stimulus, but also selectivity relative to a particular stimulus.

Since it is principally the level of the steady potential which is changed in polarization experiments, it can be concluded that a change in steady potential toward a definite optimum facilitates the manifestation of memory.

Trace phenomena in the presence of a dominant focus produced by direct current are characterized by preservation of the basic property of the focus of excitation: rhythmic nature, localization, and stability at a certain stage of development. During inhibition the imposed rhythm continues to remain in a latent state in the cortical dominant focus, and becomes manifest after inhibition.

Chapter IV

Diffuse Effects in the Central Nervous System. The Reticular Formation and the Dominant Focus

1. The Data of Classical Physiology on Diffuse Effects in the Central Nervous System

Physiologists were familiar with diffuse effects in the central nervous system at the end of last century. Vvedenskii, for instance, knew that a wave of excitation can spread diffusely over the entire central nervous system and that this spread must have definite functional significance. "From the whole nervous system," wrote Vvedenskii (1899), "a combined entity is obtained, in which the slightest change in one part is reflected more or less sensitively, at the same time or after a longer time interval, in all its other parts."*

Vvedenskii's hypothesis of the diffuse effect in the central nervous system, which was supported by a number of facts, led him to develop interesting views on intracentral relationships and on the role of extraneous central effects. Let us examine the most important of them. A state of excitation of some reflex centers is constantly echoed in the level of excitability of others; the effects are twofold: either the reflex center which is excited primarily depresses or inhibits the activity of the others or, conversely, it induces increased excitability in them, and holds them in a state of

---

*N. E. Vvedenskii. Complete Collected Works [in Russian], Vol. 6, Leningrad University Press, Leningrad (1956), p. 184.

increased readiness for excitation. Tonic excitation of some reflex centers is echoed in other centers sometimes by depression, sometimes by elevation of excitability: application of another stimulus sometimes does not paralyze an existing reflex but, on the contrary, potentiates it. If tonic excitation is produced somehow or other in both superior laryngeal nerves, moderate stimulation of the sciatic nerve gives the same effect as if the stimulus had been applied also to the superior laryngeal nerve. If both vagus nerves are brought into tonic excitation, the same stimulation of the sciatic nerve potentiates the effect of excitation of the vagus nerves (Vvedenskii, 1881). "When weak excitation arrives from a sensory nerve it may be insufficient to evoke an effect in the nearest centers, whereas as it spreads further and further, even though considerably weakened in strength, it may reach centers already in a state of excitation, and potentiate the action of these centers." In these cases Vvedenskii concentrates his attention not on nervous pathways (hence his criticism of Exner's "Bahnung"), but on the "state of activity of parts of the central nervous system — nerve cells: probably a higher level of excitability, a higher molecular mobility and state of preparedness to respond by their own reaction, is aroused in these cells."* How close these views are to that subsequently developed by Ukhtomskii in his theory of the dominant focus!

Vvedenskii and Ukhtomskii (1908) described the diffuse wave of excitation in a combined paper: Excitation arising in the central nervous system can spread extremely widely in it to its remotest parts and beyond: "In fact it must be accepted that o n e s i n g l e w a v e o f e x c i t a t i o n [emphasis mine — V.R.], reaching the central nervous system, may exhibit its action . . . on very distant centers of that system, if these centers are first prepared for this by various influences."†

It was not Ukhtomskii who first described extraneous influences in the centers, but Vvedenskii, and these pronouncements prepared the soil for Ukhtomskii's dominant.

---

*N. E. Vvedenskii, Complete Collected Works [in Russian], Vol. 6, Leningrad University Press, Leningrad (1956), p. 188.
† Ibid., Vol. 4 (1953), p. 318.

The diffuse character of excitation was brilliantly demonstrated in the 1920s by Orbeli and Kunstman (1921) on a dog after destruction of all the afferent connections of a limb. The dorsal roots of all the spinal nerves participating in the innervation of the left hind limb were divided. Among the phenomena observed after this operation, the one which attracted most attention was the regular contractions of the extensors on the left side and of the flexors on the right side, in a rhythm which coincided exactly with the rhythm of respiration, and in the phase of inspiration. Orbeli (Kunstman and Orbeli, 1924), in his analysis of this phenomenon, emphasized that after division of the sensory fibers, the limb is not paralyzed, but subjected to the influence of all reflexogenic zones, unlike under normal conditions. He was referring to a diffuse wave of excitation from the respiratory center, spreading over the entire spinal cord. Usually this diffuse wave of excitation is masked, or as Orbeli expresses it, hidden behind inhibitory influences, but in pathologically changed foci it produces the effect which corresponds to the state of the centers.

As long ago as in 1937, Beritashvili directed attention to the structure of the reticular system and postulated its role in the regulation of excitability of the brain and in particular of the spinal cord, and its role in the manifestation of "general inhibition."

Many more examples could be given to show that physiologists were familiar with diffuse effects in the central nervous system even before the discovery of the reticular system. The discovery of that system by Moruzzi and Magoun (1949) revealed the structure responsible for these diffuse effects. What is important is their relative diffuseness.

2. The Reticular Formation of the Brain Stem and "Arousal Reactions"

The inhibitory influence of the brain stem of the spinal cord was discovered by Sechenov (1863). He wrote: "The mechanisms inhibiting reflex movements lie in the thalamus and medulla."

Magoun (1944) found that certain regions of the reticular formation influence not only spinal motoneurons, but also motoneurons of the cranial nerves. Moruzzi and Magoun (1949) showed that stimulation of the reticular formation of the brain stem can cause blocking of the underlying cortical rhythm of the EEG. They explained the generalized, bilateral effect which they obtained on the EEG by the diffuse activating effect of the ascending reticular system on the cortex.

Since then facts have steadily accumulated as a result of which the view that the reticular system is a diffuse collection of cells and fibers has been questioned, and the nonspecific reticular system is now considered to be relatively independent of the specific (Filimonov, 1959; Zhukova, 1959; Brodal, 1960). Microelectrode recordings of single unit activity have revealed the detailed organization of the ascending reticular system: convergence of impulses on most reticular neurons is limited, and is obtained from different, yet extensive, receptive fields. There are inconsistencies in the literature regarding the actual definition of the reticular formation and, in particular, the structures which belong to it.

Rossi and Zanchetti (1960) emphasized that both physiological and morphological research indicates that most effects of the reticular system are not diffuse. Anokhin (1956, 1957, 1962) and his collaborators have found that the biological specificity of a stimulus (nociceptive, alimentary) is reflected in the electrical activity at the level of the reticular system, where it determines the nature of the activating influence of that system on the cortex.

The nonspecific thalamic projection system has a definite and precise organization of its cells and fibers. During electrical stimulation of the medial nuclei of the thalamus, the electrode has only to be shifted by a fraction of a millimeter for a different response to be obtained in the cortex (Jasper, 1960; Pavlygina, 1962). In Jasper's opinion, the thalamic reticular system plays an important role in the distribution of the level of excitability between different cortical regions.

That the arousal reaction is obtained not only to stimulation of the reticular formation, but also to stimulation of other brain structures such as the amygdala (Feindel and Gloor, 1954), the head of the caudate nucleus (Stoupel and Terzuolo, 1954), and certain structures of the cerebellum is clear from the increasing volume of facts reported in the literature.

Diffuse effects can also arise in the EEG from a stimulus applied to the cortex itself, and they persist for a long time after division of the white matter and corpus callosum (Sloan, 1950; Bremer, 1954; Ricci, 1955). These experiments show that the transmission of stimulation and the pathways of spread of excitation evoking the arousal reaction pass both through the reticular system and also, bypassing it, through the specific system.

An arousal effect can be obtained after destruction of the mesencephalic reticular formation, when all the classical pathways are left intact, but under these circumstances there is no long aftereffect. The limited dependence of the arousal reaction on the reticular formation has been shown experimentally.

Both the electrographic reflection of the arousal reaction itself and the functional state of the cortex during this reaction require further analysis with the use of mathematical methods. Microelectrode studies have shown that during inhibition of the underlying EEG rhythms some neurons are excited while others are inhibited. It is therefore not yet clear with what state of the neural substrate the activating reaction must be related.

The systems of neurons whose direct electrical stimulation evokes a generalized arousal reaction is itself under the influence of impulses descending from the cortex. Conduction of afferent impulses not only in specific relay nuclei, but also in nonspecific structures can be regulated by the cortex. In nearly every case changes in evoked potentials observed under the influence of stimulation of the reticular formation could also be evoked by cortical stimulation. Hernández-Péon and Donosso (1959) investigated evoked potentials in the optic radiation of patients and found that concentration of attention on the stimulus concerned is an important factor in the manifestation of the responses. They consider that the cortex activates the reticular formation which, in turn, influences the specific visual pathways. However, the hypothesis that the flow of afferent impulses is regulated entirely by the reticular formation cannot explain all the facts known to physiology of the central nervous system. There are direct corticofugal pathways which influence relays of the specific afferent system.

Filimonov (1959) emphasized that the reticular formation is a derivative of the visceral tube, i.e., that it shares a common origin with the intermediate zone of the spinal cord. Evidence in support of this view is given by their structural similarity and also by the link between both and autonomic functions. Klosovskii (1959) also maintains that the cell concentrations of the reticular formation are "autonomic centers." They are undoubtedly participants in reflex activity, although not by direct activation of the cortex but through provision of an autonomic component characteristic of all brain activity.

Bonvallet, Dell, and Hiebel (1954) showed that nociceptive stimulation in animals evokes prolonged cortical activation, accompanied by an increase in sympathetic tone. Cortical activation is produced by the action of adrenalin via structures of the mesencephalon, but not by its direct action on the cortex. These results were later confirmed and analyzed by Rothballer (1956) and Karamyan (1959).

Karamyan (1962) regards interaction between nonspecific and specific brain systems in the light of Orbeli's theory of functional evolution. After the work of Dell and his collaborators, investigations began into the role of adrenalin as an exciter of the adrenergic substrate of the reticular formation.

Investigations of the mesencephalic reticular formation and its complex interaction with the cortex, together with microelectrode analysis of unit activity at different levels of the central nervous system have demonstrated the wide possibilities of convergence of afferent impulses on single units. Such convergence may be of two types: one reinforced by existing anatomical connections; the other functional, dynamic, and formed in the course of a reflex response upon the appearance of foci of excitation. The second type, in particular, arises in the presence of a dominant focus.

3. Similarity between EEG of the Motor and Visual Cortex in the Presence of a Motor Dominant Reinforced by Photic Stimulation

The problem of foci of excitation acquiring the characteristics of a dominant, when the same reflex response is obtained regardless of the site or type of stimulation, is closely bound up with the view of the diffuse spread of excitation in the central nervous system and, at the same time, of its local manifestation. Without the concept of the diffuse spread of excitation in the central nervous system and the presence of physiological mechanisms for the functional, dynamic convergence of impulses of different modalities, the theory of the dominant could not have been established.

Sokolova (1958) investigated electrical activity of the cortex and of certain subcortical structures (lateral geniculate body, caudate nucleus) of the rabbit, producing a cortical dominant focus by weak direct current. To obtain a motor response to photic stimulation, an optimal level of excitation had to be created not only in the motor, but also in the visual system. Besides optimal strength of the current polarizing the motor cortex, optimal strength of photic stimulation also was needed. Under these conditions, in response to photic stimulation an isolated movement of the limb contralateral to the polarized area of cortex was obtained.

When a limb movement is obtained in response to photic stimulation, the EEGs of the visual and motor areas appear similar. This may be reflected in coincidence of the phases of depression and exaltation of electrical activity, or of an increase in frequency of the waves on the visual cortical EEG. In the latter case, the EEG of the visual cortex came to resemble the EEG of the motor cortex in its external appearance. The greater the similarity between electrical activity of the motor and visual areas, the easier it was to obtain a motor response to photic stimulation during cortical polarization.

The increasing similarity between EEGs of the visual and motor areas and coincidence of the phases of depression and exaltation are evidence, from the point of view of types of functional interneuronal communication, of identical phases of the electrotonic states of these areas, reflected in their EEGs. The electrotonic effects evidently ensure identity of functional state of the two cortical areas, which is essential for conditioning. Functional unity of the two cortical areas is reflected not only in the similarity between their electrical activity, but also in their responses to afferent stimuli: the motor area becomes able to respond to interrupted photic stimulation at the same time as the visual area. The observed similarity in electrical activity of the visual and motor cortex is evidently an electrographic reflection of the synthesis of neural processes during the formation of temporary connections, as established by Pavlov (1926), Kupalov (1947), Voronin (1957), and others. The similarity between the electrical activity of the visual and motor areas has also been observed in experiments with the dominant focus, particularly in intervals between stimulation. It was clearest at moments preceding limb movement in response to photic stimulation.

Consequently, the special features of the response resulting from formation of a temporary connection in the presence of a motor dominant focus are not revealed actually during the reflex movement, but before it, and they take the form of a change in electrical activity of the cortical areas of the connected systems. At the actual moment that the reflex movements take place, the changes in the visual and motor areas are proceeding in different directions: in the motor cortex electrical activity is falling, and against its background fast waves are appearing, while in the visual cortex a rhythm of 5/sec is recorded.

At the end of the motor response, the electrical activity of the visual and motor areas resumes its previous appearance: the unity of these areas, as shown by their local electrical response, is restored. The pattern of electrical activity during limb movement in response to photic stimulation as described above is observed at optimal strengths of the direct current (1-5 μA) which do not significantly change the electrical activity of the motor cortex. If, however, the strength of the direct current is increased, and significant changes take place in the EEG of the motor cortex, the relationship between the electrical activities of the visual and motor areas also changes. Whereas waves of higher amplitude or higher frequency than in the original record appear in the motor area, activity of the visual area shows little change from its original form. During the limb movement depression is observed in the EEG of the visual cortex and an increase in electrical activity (an increase in amplitude of the waves) in the motor area. In these cases, interrupted photic stimuli, which are known to be stronger than single stimuli, usually have to be applied in order to evoke movements. The changed electrical activity of the motor cortex is evidently not a favorable background for onset of the motor response to afferent stimulation. As has already been emphasized, electrical activity during which there is merely a certain increase in the number of fast waves compared with the original record must be regarded as the optimal level of electrical activity. On the other hand, for a motor response to develop to photic stimulation, the optimal conditions are those under which the EEGs of the cortical areas are similar.

4. Electrical Activity of the Lateral Geniculate Body and Caudate Nucleus in the Presence of a Cortical Dominant Focus

The electrical activity of the lateral geniculate body and caudate nucleus has been investigated during motor responses to photic stimulation occurring in the rabbit during cortical polarization by a weak direct current (Sokolova, 1958). In the original records great similarity was seen between the electrical activity of the lateral geniculate body and of the visual cortex. This similarity was still present even when the EEG of the motor cortex was modified after dc polarization. The electrical activity of the lateral geniculate body and visual cortex were similar also when the focus in the motor area was reinforced by photic stimuli. With the onset of a limb movement in response to light, as has already been said, the EEGs of the visual and motor areas became similar. The electrical activity of the lateral geniculate body changed parallel with that of the visual cortex. As a result, the electrogram of the lateral geniculate body, like that of the visual cortex, came to resemble in its background activity the EEG of the motor cortex. When the EEG of the motor area followed the rhythm of the flashes in response to interrupted photic stimulation, a rhythm-driving response was also observed in the lateral geniculate body and visual cortex.

The similarity found in the EEGs of the visual and motor cortex on reinforcement of a dominant focus in the motor cortex by photic stimulation thus extends also to the electrical activity of the lateral geniculate body. When the EEG of the motor cortex is considerably changed by the direct current, the electrical activity of the lateral geniculate body remains similar to that of the visual cortex. Consequently, during polarization of the motor cortex, the visual cortex and lateral geniculate body form a single system with the motor cortex. Only if the polarizing current and photic stimuli are optimal does a reflex movement of the corresponding limb develop in response to photic stimulation.

During her investigation of electrical activity of the caudate nucleus, Sokolova compared it constantly with the EEG of the motor cortex. She found that in the initial records the activity of the caudate nucleus was similar to the EEG of the motor cortex. However, this similarity may appear and disappear in the course of the same experiment. The motor cortex evidently influences the caudate nucleus, after which their electrical activities are synchronized.

The influence of the motor cortex on electrical activity of the caudate nucleus is clearly seen during the action of direct current on the cortex. If the current is strong, and electrical waves appear in the cortex in the rhythm of respiration, this same activity is observed after a time in the caudate nucleus. The influence of the cortex on electrical activity of the caudate nucleus is observed when the dominant focus in the motor cortex is reinforced by photic stimulation. In these cases the unity of electrical activity in the different parts of the brain extends to the caudate nucleus to a varied degree, depending on the stage of development of the dominant focus and on the strength of the cortical influence on the caudate nucleus. However, even when the electrical patterns of the motor cortex and caudate nuclei were most similar, activity in the caudate nucleus never followed the rhythm of the flashes, as it did simultaneously in the motor and visual areas during reflex movements of the limb to photic stimulation. Hence, although the caudate nucleus is united in its local electrical activity with the visual and motor areas during reinforcement of a motor dominant focus by photic stimulation, its electrical response to photic stimulation does not come to resemble the visual cortex, as does the motor cortex.

Sokolova's results thus show that the caudate nucleus does not necessarily take part in the formation of a single functional system in the visual and motor systems, but does so only insofar as it is under the influence of the motor cortex. Connections between the cortex and basal ganglia have often been investigated. In particular, Glees (1944) showed that the "areas of inhibition" in the cat's cortex are connected with the caudate nucleus by unmyelinated fibers, which are collaterals of corticofugal fibers. It also follows from the work of Dusser de Barenne and McCulloch (1938) that connections exist between the cortex and caudate nucleus. Others believe that the systems of the corpus striatum, consisting of caudate nucleus, putamen, and globus pallidus, have no direct connections with the cortex. However, it has recently been shown that during repetitive electrical stimulation of the caudate nucleus, a "recruiting response" appears in the motor cortex, indicating a retrograde connection between caudate nucleus and cortex.

An important investigation into the function of the caudate nuclei was undertaken by Klosovskii and Volzhina (1956). To study the function of the caudate nucleus they extirpated it bilaterally, in one stage, in puppies aged 2-3 months. During the first day after the operation they observed an alternation of periods of marked excitation and rest. During excitation the puppies were in continuous movement. They did not respond even to strong photic or acoustic stimulation. On the day after the bilateral extirpation most of the puppies could stand and walk. The tone of the fore- and hind limbs was equal. These puppies slept so deeply that external stimulation awakened them only with difficulty. The transition from sleep to waking was not instantaneous, as in normal puppies, but very slow.

The behavior of dogs after extirpation of the caudate nuclei at the age of 2-3 months was observed for 2-3 years. Bilateral extirpation has a marked effect on CR activity. Either CRs could not be formed in these animals, or they were formed very slowly and were unstable. Klosovskii and Volzhina concluded from their experiments with extirpation and division of various structures that bilateral extirpation of the caudate nuclei or division of the tracts from these structures to the thalamus completely disrupts behavior and disturbs neural activity despite the generally satisfactory somatic state of the animals. Some degree of compensation takes place later, but the formation of CRs remains severely disturbed.

It is interesting to note that changes in cortical and subcortical electrical activity obtained in acute experiments by Sokolova in the presence of a dominant focus in the motor cortex, resulting from polarization by a weak direct current, resemble the EEG changes found by Livanov, Korol'kova, and Frenkel' (1951) during conditioning, especially in the generalization phase. This similarity between phenomena with respect to their motor effect and the changes in their electrical activity obtained during the formation of a dominant focus in the motor cortex and during the initial phase of motor conditioning is evidence, not only of the close similarity, if not identity, between the physiological mechanisms lying at their basis, but in my opinion, it is also evidence that electrotonic influences are essential for formation of temporary connections. In all experiments the dominant focus was evoked by a weak direct current, similar in its action to the electrotonic effect.

5. The Cortical Dominant Focus in Long-Term Experiments

Sokolova and Khon Sek Bu (1957) formed a focus of excitation in the motor cortex in long-term experiments on rabbits with implanted electrodes and observed the course of changes in cortical electrical activity. When the active electrode for polarization of the motor cortex is an anode (reference electrode on the rabbit's ear), an increase in frequency of the EEG is observed in the polarized part of the cortex. Within a few minutes after onset of polarization, well-defined fast waves with a frequency of 40-44/sec and amplitude up to 50 μV appear in the polarized area. As the current continues, the amplitude of these waves increases still more — to 70-100 μV.

The increased frequency of the electrical activity spreads to neighboring areas and to the sensorimotor area of the opposite hemisphere. After removal of the direct current, the fast waves continue from a few minutes up to 1-2 h. With cathodal current applied to the same region, fast waves also appear. However, this rhythm develops differently from that with anodal current. Initially, slow waves of high amplitude and long duration appear. With continuation of the cathodal current these slow waves are reduced, the fast waves against their background are increased, and within a short time the EEG of the polarized area begins to be dominated by faster waves (about 40/sec), similar to those observed during anodal current.

During polarization of the motor cortex with implanted electrodes, just as in acute experiments, a reflex motor response to

5. THE CORTICAL DOMINANT FOCUS IN LONG-TERM EXPERIMENTS

hitherto indifferent stimuli is observed. In the clearest cases this response is expressed as movement of the rabbit's forelimb contralateral to the polarized area.

The optimal strength of the direct current in the experiments of Sokolova and Khon Sek Bu was between 3 and 5 μA. The motor response to hitherto indifferent stimuli appeared 10-15 min after the beginning of polarization.

During the motor response, a fast rhythm appeared in the EEG of the sensorimotor cortex, or if fast waves had already appeared in the cortical EEG under the influence of the direct current, their amplitude was increased. In some experiments, they observed the appearance of slow waves of high amplitude during movement of the forelimb, but later in the same experiments these waves could disappear and be replaced by slowing of the rhythm during one of the subsequent movements. These slow waves are possibly similar to those described by Novikova et al. (1952).

The appearance of a motor response to a hitherto indifferent stimulus during polarization of the motor cortex takes place gradually, and other conditions being the same, it depends on the strength of the stimulus. The response appears initially to "stronger" stimuli (acoustic), and then to "weaker" stimuli (photic).

In long-term experiments on rabbits, the dominant focus in the motor cortex did not disappear immediately after removal of the direct current, but persisted for a few minutes to 2-3 hours in different experiments. As a rule, the length of time that the motor effect persisted after removal of the direct current depended on the intensity of the dominant focus. When the dominant focus reached its highest level of development, the motor reflex continued for 2-3 h in the afterperiod. When the dominant focus was weaker, the motor response to afferent stimulation disappeared more rapidly after cessation of polarization. In short-term experiments the longest duration of persistence of the dominant focus, as mentioned above, did not exceed 20-40 min.

Persistence of the dominant focus after removal of the current evidently reflects inertia of the stationary excitation created in the cortex by dc polarization. This inertia is also reflected in the EEG. Quickening of the waves in the sensorimotor cortex at an anode continues for various times after removal of the current, also depending on the intensity of the dominant focus.

When a dominant focus in the sensorimotor area for some reason or other was not formed by the polarization, an increase in the frequency of the EEG could still arise in the polarized region. However, this increase in frequency disappeared very quickly after removal of the current.

Thus, the stability of preservation of the faster rhythm evoked by the anodal polarization is intimately linked with the stability of the dominant focus in the sensorimotor area. In the experiments of Sokolova and her collaborators, just as in the short-term experiments, polarization of the rabbit cortex by weak direct currents through implanted electrodes produced all the basic features of the dominant focus, especially its stability and inertia, i.e., it continued after cessation of the current which produced the focus itself.

6. Electrical Activity of the Medial Geniculate Body and Caudate Nucleus in the Presence of a Motor Dominant Reinforced by Acoustic Stimulation

Changes in the electrical activity of the auditory cortex, the caudate nucleus, and the medial geniculate body during anodal polarization of the motor cortex were investigated by Naumova (1953, 1956). Like other investigators, she found that changes in electrical activity in the zone of polarization and adjacent areas were closely dependent on the conditions of polarization and the state of the polarized zone. If weak currents of threshold strength were used, anelectrotonus was associated with an increase in frequency of the waves to 15-19/sec. If polarizing currents of minimal intensity, of optimal strength for creation of the dominant focus, were used there was a significant difference between the action of the anode and cathode on processes reflected in the EEG during polarization. Usually under these conditions the anode caused the appearance of faster waves. The cathode left the electrical activity of the polarized zone unchanged. During polarization by a stronger current, the action of the anode, like that of the cathode, was accompanied by an increase in frequency, sometimes accompanied by the appearance of cardiac and respiratory rhythms in the EEG. The functional state of the motor cortex in which a motor reflex took place in response to acoustic stimulation was reflected electrographically by an increase in frequency. According to Naumova, movement of the limb in response to acoustic stimulation appeared only when the stimulus evoked an additional, but moderate increase in frequency of the electrical waves.

As was mentioned above, the motor response to acoustic stimulation occurred against the background of waves faster than before polarization. This increase in frequency of the cortical electrical oscillations was observed only in the zone of anelectrotonus with minimal strengths of polarizing current.

The reasons why the anode was more effective, as Fritsch and Hitzig (1870) observed orginally during cortical stimulation, may be that the cortex has different functional mobility from that of nerve, and this determines the "reverse" response of the cortex to application of the dc poles. The appearance of a motor response to acoustic stimulation against the background of weak anelectrotonus accords with Udel'nov's observations (1938) of the superiority of the anode for the formation of a single tetanic contraction, regarded as the prototype of the dominant.

In most experiments the electrical activity of symmetrically opposite motor areas was identical during the motor response. It was identical when the effector producing the response was that whose cortical representation was subjected to dc polarization. The appearance of identical changes in symmetrical zones of the motor areas during the motor response in the presence of the dominant focus is evidence that equivalent changes of functional state take place in these zones. However, despite the identity of the EEGs in symmetrical motor areas, the limb whose cortical representation is being directly polarized makes the greatest movement to acoustic stimuli.

Acoustic stimulation in the presence of a dominant focus, as mentioned above, led to an additional increase in frequency of the electrical activity at the site of polarization of the motor cortex. At the same time, a decrease in the amplitudes and changes in frequency were observed in the temporal and parietal regions. Simultaneous changes in electrical activity arising in different parts of the cortex during acoustic stimulation are evidently produced by electrotonic effects. The electrical nature of these effects is shown by the fact that responses in different parts of the motor area and parietal region to acoustic stimulation are mutually related: they rise and fall in unison. While the motor response to an acoustic stimulus is taking place, a rhythm of 5-6/sec frequently appears in the parietal and temporal regions.

It is interesting to note that if the acoustic stimuli are applied at a definite interval, when the stimulation ceases, waves similar to the responses observed during actual acoustic stimulation still appear at the same time intervals.

The electrical activity of the caudate nucleus and medial geniculate body also changes during the formation of a dominant focus in the motor cortex by dc polarization. Study of the caudate nucleus was undertaken since among the subcortical structures belonging to the motor system it is nearest to the cortex. Changes in electrical activity were examined during acoustic stimulation before polarization, against the background of the focus produced by polarization, and in the period immediately after removal of the direct current. At weak intensities of cortical polarization no change could be found in the electrical activity of the caudate nucleus. A stronger direct current led to the appearance of faster waves than those observed before polarization. The phenomenon thus described occurred when electrical activity of the caudate nucleus before polarization was similar to the cortical electrical activity of the same animal. The tips of the recording electrodes were in the middle part of the head of the left caudate nucleus.

The electrographic pattern of activity of the caudate nucleus accompanying the motor response of the limb to acoustic stimulation was indistinguishable from the pattern observed during application of the acoustic stimulus before polarization. A motor response occurred only if the acoustic stimuli, just as before polarization, evoked a decrease in amplitude and change in frequency of the electrical activity in the cortex and caudate nucleus. In those cases where the activity of the caudate nucleus was considerably altered by comparison with the background, or when synchronized slow waves occurred in the cortex and caudate nucleus, no motor response took place to the acoustic stimulus. If the background activity was similar in the cortex and caudate nucleus, the presence or absence of a motor response depended on the action of the acoustic stimulus on the motor cortex. If the stimulus produced no changes in the EEG of the polarized region, there was no motor response. If, however, the acoustic stimulus caused depression in the cortical EEG, movement of the limb ensued. Changes arising in the caudate nucleus in the presence of a dominant focus differed very little from those evoked by acoustic stimulation before polarization. These changes took the form of a decrease in amplitude and increase in frequency of the activity, thus indicating its similarity with the changes in the motor area.

Changes in the electrical activity of the medial geniculate body were similar to those appearing to acoustic stimulation before polarization. In both the magnocellular and parvocellular nuclei of the medial geniculate body, both of which form part of the auditory system, acoustic stimulation either depressed the electrical activity or produced waves at the onset and end of stimulation.

The main conclusion from Naumova's investigations is thus: when a focus of excitation is produced in the motor cortex, the changes in electrical activity which take place in response to acoustic stimulation differ from those evoked by the same stimulation prior to polarization. Consequently, the decisive factor for the appearance of a motor response to acoustic stimulation is the processes taking place in the cortical dominant focus itself.

Do pathological foci in patients, especially those with brain tumors, possess the properties of dominant foci?

Grindel' and Filippycheva (1959) studied disturbances of human motor activity in patients with a pathological focus (tumor) in the frontal lobes and demonstrated the presence of excitation of a stationary character, resembling a pathological dominant focus in the affected motor area. Their investigation involved recording the conditioned motor response. They found that the rhythmic character of excitation of the actual focus is just as clearly revealed as in experiments on animals. After a series of stimuli, a repetitive response is obtained even to a single stimulus. Perseveration has long been known as a symptom of frontal lesions in man. Its physiological mechanism is evidently the rhythmic character of the state of stationary excitation in the focus, which thus acquires the properties of a dominant. The work of Grindel' et al. shows that if a patient's rhythmic motor response to acoustic stimulation begins to disappear, any indifferent stimulus (photic, tactile) is sufficient to cause the amplitude of the motor responses to increase once again, although their rhythm remains as before.

These results suggest that the pathological focus in some cases of organic brain lesions in man is similar to a dominant focus. Electrophysiological investigations show that the pathological focus is a source of dynamic changes of state over the entire cortex, so that its "focal isolation" becomes exceedingly relative.

Electrocorticographic investigations during operations on the exposed human brain show that electrical activity changes a very short distance (1-2 mm) away from the focus. Afferent stimulation applied during the operation emphasizes the differences even more (Maiorchik, 1964).

Chapter V

Cortical—Subcortical Relationships, Thalamo-Cortical Connections, and Dominant Foci in the Hypothalamus

1. Polarization of Thalamic Areas

In the course of investigation of cortical—subcortical relationships in the presence of a dominant focus the question arose as to the effect of the diffuse thalamic system on an experimentally created focus of excitation in the cortex. As Pavlygina (1960) showed, additional polarization of the medial nuclei of the thalamus by a weak direct current against the background of an existing cortical dominant considerably increases the movement evoked by indifferent stimuli (Fig. 25). The same effect on a cortical dominant focus was found by Kalinin (1965) during additional polarization of the mesencephalic reticular formation by a weak direct current in rabbits. Increasing the strength of the polarizing current applied either to the medial thalamic nuclei or the mesencephalic reticular formation at first strengthens the dominant focus, but later inhibits it.

Kalinin investigated the effect of polarization of the specific thalamic nuclei (the medial and lateral geniculate bodies) on a cortical dominant created by weak polarization of the motor cortex. He found that polarization of the thalamic relays reinforces the dominant only if the polarized thalamic nucleus participates in the

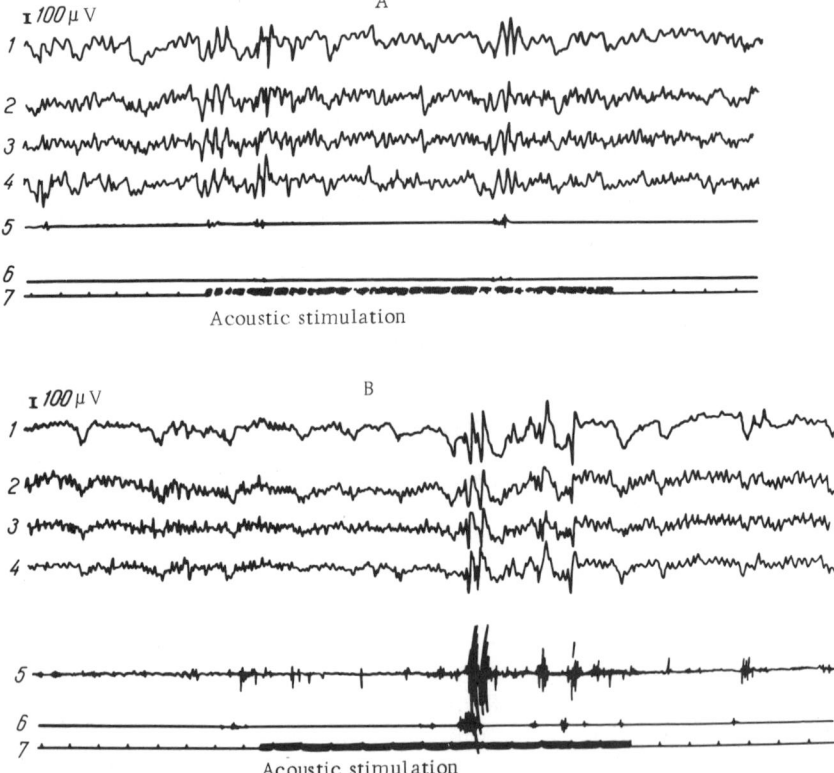

Fig. 25. Marked increase in response to afferent stimuli during additional polarization of medial thalamic nuclei by a weak direct current. Formation of two foci of excitation by a pulsating current (work of Pavlygina). A) pulsating current on motor cortex (constant components 0.1 µA, variable 0.2 µA, frequency 3 Hz); B) additional action of same current (from a different apparatus) on medial nuclei of the thalamus; increased limb movement. 1) right visual area; 2) right motor area; 3) left motor area; 4) right reticular formation; 5) movements of left forelimb; 6) movements of right forelimb; 7) time (in sec) and acoustic stimulation.

transmission of impulses from the stimuli applied. When polarizing the motor cortex, like other investigators before him, Kalinin obtained a motor reflex of the limb corresponding to the polarized cortical point in response to hitherto indifferent acoustic and photic stimuli. If, in addition, he polarized the lateral geniculate body

1. POLARIZATION OF THALAMIC AREAS

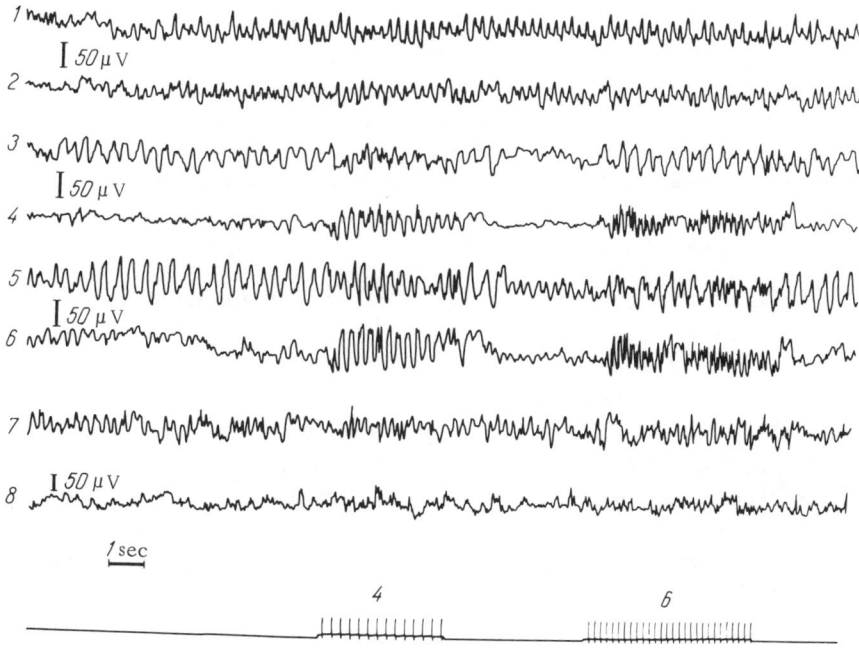

Fig. 26. Asymmetry of electrical activity recorded from the cortex of a rabbit during polarization of the lateral geniculate body (dc, 5 μA) (work of Kalinin). 1) right motor area; 2) left motor area; 3) right parietal region; 4) left parietal region; 5) right visual area; 6) left visual area; 7) right medial geniculate body; 8) right lateral geniculate body. Bottom line: repetitive photic stimulus.

by a weak direct current, he obtained strengthening of the dominant focus only to photic stimulation. To strengthen the dominant, i.e., to increase its stability, to increase the number of motor responses of the limb, and to increase their strength in response to acoustic stimuli, the medial geniculate body also has to be additionally polarized by a weak direct current. Kalinin investigated electrographically the effect of additional polarization of the lateral geniculate body and showed that the following of the rhythm of photic stimuli also is improved. The polarizing current has its own optimum of action on the thalamic nuclei just as on other structures of the central and peripheral nervous systems. Prolonged polarization or an increase of the direct current does not reinforce the dominant, but inhibits it. For example, an increase in the current from

1-3 to 5-6 μA, when acting on the lateral geniculate body, inhibits the dominant focus: no movements arise in response to photic stimulation, whereas the same dominant focus continues to respond as before to acoustic stimulation, i.e., a motor response to acoustic stimulation is obtained. Results of a similar pattern are obtained in the medial geniculate body if the strength of the polarizing current is increased: the response to acoustic stimulation is inhibited while that to photic stimulation continues.

When a cortical dominant focus is present, the predominant rhythm in the subcortical structures is usually 4-6/sec. Increasing the current to 5-10 μA or more for subcortical polarization leads to the appearance of slow waves in these structures, which then appear in the surface EEG. Asymmetry of the hemispheres becomes apparent in the cortical electrical activity (Fig. 26). This figure shows that slow, high-amplitude waves appear in the visual and parietal region of the cortex on the right side, after polarization of the right lateral geniculate body for 25 min. Asymmetry between the hemispheres is clearly visible. The cortex of the right occipital and parietal regions responds with a rhythm of 3-4/sec to a repetitive photic stimulus (6 flashes per second). This is the only stimulus which reinforces the dominant focus, the only one which can create the necessary and sufficient level of excitation in the corresponding thalamic relays. This is one of the reasons why a motor response of the limb does not always arise, and does not arise to every frequency or intensity of the applied stimulus, during polarization of the sensorimotor cortex.

To form a dominant focus, when a cortical focus of excitation is present, the optimal level of excitation is required not only in the cortical focus itself, but also in the thalamic relay of the system transmitting the afferent stimulation.

Intracentral relationships leading to the formation of a dominant are only one aspect of these complex interactions. It is not merely that the cells of the cerebral cortex respond differently to the same external stimulus and, consequently, to the same afferent volley of impulses, depending on their functional state. A focus of excitation formed in the cortical representation of a sensory system leads to an essential change in the structure of the afferent volley itself, depending on the state of that focus. The work of Meshcherskii (1955) showed that after combination of conditional and unconditional stimuli, a state similar to that of the dominant focus arises in the cortical area of the system receiving the CS. In their investigation of some of the more complex physiological mechanisms lying at the basis of this phenomenon, Meshcherskii and Gustson (1964) showed that during unilateral depression, spreading over the rabbit's cortex, the amplitude of the primary responses of the ipsilateral lateral geniculate body to flashes is reduced, and the positive component observed in the responses of the lateral geniculate body is completely blocked by strychninization of the visual cortex.

These findings are evidence that the cortex exerts local control, by specific corticofugal pathways, over the thalamic relays as the result of recurrent corticothalamic connections.

2. Polarization of the Medial Nuclei of the Thalamus

Does a weak direct current differ in its action on the specific and nonspecific nuclei of the thalamus? In other words, if the specific system has its own optimal strength of direct current, and its own strength of current producing the corresponding depression, are these also found for the nonspecific, diffuse system of the thalamus?

In her investigation of the action of direct current on the medial thalamic nuclei of rabbits in long-term experiments, Pavlygina found the following interesting behavior with respect to rhythmic driving by flashes during weak, anodal polarization. In response to flashes at 2/sec, a rhythm of 4/sec appeared. The results given by Walter's frequency analyzer show that the amplitude of the rhythm at 2/sec is not increased, but the amplitude of the rhythm at 4/sec is considerably increased. After removal of the photic stimulation the amplitude of the rhythm at 4/sec falls to its initial value, but the amplitude of the rhythm at 2/sec is increased. If the current polarizing the medial thalamic nuclei is increased, the phenomenon of driving disappears.

During polarization of the mesencephalic reticular system, a generalized change in electrical activity is observed in the cortex and thalamus. Kalinin polarized the mesencephalic reticular formation of a rabbit and obtained results indicating the diffuse effect of such polarization, although the effect was more marked in the visual and parietal regions of the cortex and at the point of polarization. The changes subsequently irradiated from the occipital regions to the anterior regions of the brain. During more prolonged polarization of the mesencephalic reticular system by a stronger current (up to 20 μA), generalized depression of electrical activity takes place in the cortex and the medial and lateral geniculate bodies, and the limit of the frequency of flashes which can be followed is lowered.

These facts show that there is an optimal strength of the direct current, as well as a strength producing depression, both for the specific and for the reticular system, although there is a substantial difference in their ultimate effect on the cortex. Polarization of the specific thalamic nuclei does not give rise to diffuse changes in the cortex.

3. Polarization of the Hypothalamus

Investigation of a focus of excitation produced in the hypothalamic region by a weak direct current is of considerable interest. Stimulation of the hypothalamus leads to changes in respiration, cardiac activity, and metabolism.

Von Euler (1950) recorded slow potentials in the cat's hypothalamus. These were evoked by raising the brain temperature artificially and were recorded locally only from the anterior hypothalamus. The slow changes in potential corresponded precisely to the ensuing thermal tachypnea and tachycardia.

Pavlygina (1956) investigated dominant foci produced in the hypothalamus in rabbits by direct current in long-term experiments. She recorded the hypothalamic electrical activity at the point of application of the polarizing electrode, the EEG of the lateral surface of the cortex, the EKG, pneumogram and the blood pressure.

During the first few seconds the weak direct current evoked a definite increase in amplitude of the hypothalamic electrical activity. The respiratory rate was increased and in some cases the heart rate also changed. Characteristic changes in respiration and cardiac activity occurred during the first 10-60 sec of action of the direct current. Later, despite continuing polarization, respiration and cardiac activity returned to normal, although the amplitude of the hypothalamic electrical activity remained increased throughout the experiment.

After the initial autonomic responses had been restored, acoustic and photic stimuli to which the orienting reflex had first been extinguished evoked changes in respiration and cardiac activity which were not observed before application of the direct current. This indicated that the focus of excitation produced in the hypothalamus by the direct current possessed the basic property of a dominant. Certain conditions were essential for the creation of the hypothalamic focus. The strength of the current, the duration of its action, the strength and frequency of the reinforcing stimuli, and the state of the animal itself were all very important. To produce a dominant focus in the hypothalamus, the optimal current is from 3 to 10 μA.

It is interesting to note that at the beginning of the formation of a dominant focus in the hypothalamic region changes in cardiac activity take place in the period after photic and acoustic stimulation. As the dominant focus becomes stabilized these changes move closer to the time of the stimulus. The changes in respiration in most cases occur to acoustic or photic stimuli from the very beginning of formation of the focus. The neural systems participating in the regulation of respiration and cardiac activity exhibit different functional mobility. By forming a dominant focus by polarization in the hypothalamus, as in Magnitskii's (1952) experiments, the blood pressure of an animal can be altered.

In Pavlygina's experiments the blood pressure rose slowly throughout the experiment during polarization of the hypothalamus. Acoustic and photic stimuli applied during polarization caused no definite changes in the blood pressure. The phenomena associated with the dominant focus did not disappear immediately on removal of the direct current in all the rabbits, but continued for several hours, and in some cases actually were still exhibited for several days.

With reinforcement of the dominant focus by photic or acoustic stimuli hypothalamic electrical activity is altered and varies with the state of the focus. In the first stage of formation of the dominant focus, before its presence is clearly defined, its reinforcement by afferent stimulation is accompanied by slowing of activity in the hypothalamus, bringing it closer to the rhythm of respiration. With stabilization of the dominant the amplitude of the hypothalamic waves increase, and slower waves become predominant. If stimulation is applied against this background, even slower waves of high amplitude, 1-2/sec, appear. Simultaneously the respiratory rate increases. With photic or acoustic stimulation the amplitude of cortical electrical activity also changes. With cessation of the afferent stimulation the original amplitude of the cortical electrical activity is restored.

The more marked the dominant focus during hypothalamic polarization, the greater the decrease in amplitude of the cortical EEG upon reinforcement of the focus by afferent stimuli. Conversely, in the absence of the dominant, i.e., if no changes in autonomic responses are observed in response to stimulation, there will be no decrease in amplitude of the EEG for these afferent stimuli. In some experiments, when trace phenomena associated with the dominant focus were present, i.e., after removal of the direct current, high-amplitude slow waves with a frequency of 3-5/sec appeared in the cortex at the time of application of the afferent stimuli.

Those same factors which produce a dominant focus can, if their strength is increased, cause its inhibition. Conversion of a focus of excitation in the hypothalamus into a state of depression was observed when the direct current was increased. If the optimal current for producing a dominant focus in the hypothalamus was ≤ 5 μA, an increase to 10 μA or above led to the appearance of large, slow waves in the hypothalamus, more than 1 sec in duration. Against this background, the respiratory rate increased still further in response to photic and acoustic stimulation. If such current was prolonged up to 15-20 min, large slow waves appeared not only in the hypothalamus, but also in the cortex, first in the visual area, then in the motor area. If this higher current was continued, the slow waves in the cortex disappeared and electrical activity was sharply reduced. The slow waves in the hypothalamus, however, still persisted at that time and respiration became shallow. Under these conditions neither acoustic nor photic stimulation brought out the dominant properties of the focus.

3. POLARIZATION OF THE HYPOTHALAMUS

The fact that electrical activity is reduced in the cortex before the hypothalamus, where the current is applied, can evidently be explained by assuming that the cortex is the more reactive structure and that it can pass into a state of deep inhibition sooner than can the hypothalamus. With continued action of the direct current on the hypothalamus, electrical activity of the cortex and hypothalamus become similar, and a sharp decrease in amplitude of the EEG is observed in both loci.

Electrical activity was restored in the cortex and hypothalamus 40-50 min after removal of the direct current, bringing the dominant focus into a state of inhibition. The gradual recovery of the dominant focus from its depression was studied. At a certain stage of recovery, a regular rhythm of 5-7/sec characteristically was predominant. Just as during extinction of the orienting reflex, during deeper inhibition there was a sharp decrease in electrical activity and the almost total absence of potentials was observed. With less deep inhibition, a regular rhythm of 5-8/sec appeared.

Investigation of the dominant focus in the hypothalamus thus shows that after application of direct current changes in respiration and in cardiac activity take place in rabbits. These changes soon pass, and they are then observed later in response to photic and acoustic stimulation. Since the orienting reflex to acoustic and photic stimulation had first been extinguished, the appearance of autonomic responses to these stimuli afresh against the background of hypothalamic polarization provides indisputable evidence of the formation of a dominant focus at the site of polarization.

It might thus be suspected that disinhibition of the orienting reflex had occurred, but that this is not so is shown by the following facts. Before a change in the autonomic responses could be obtained, not one but several photic or acoustic stimuli had to be applied, and during the action of those stimuli the autonomic response increased steadily in intensity. This last fact indicates the role of summation of incoming waves of excitation in the focus created by polarization in the hypothalamus. If a weak direct current affected the extinguished orienting reflex as a disinhibiting agent, the change in autonomic responses to photic and acoustic stimuli should be observed at the onset of the direct current, rather than after adaptation to the particular stimulus. Furthermore, whereas in the orienting reflex mainly changes in respiration are observed, a change in the pulse rate and, still more, a change in the blood pressure are rarely observed, while with formation of a dominant focus in the hypothalamus both respiratory and cardiac activity are altered.

There is also a significant difference in the electrical response of the hypothalamus to acoustic and photic stimulation for the orienting reflex versus that seen in the presence of a dominant focus. In the orienting reflex, a decrease in amplitude of the electrical activity is observed in the hypothalamus, just as in the cortex, and simultaneously a regular rhythm appears. In the presence of a dominant focus, on the other hand, at the moment of appearance of autonomic responses to indifferent stimuli, slow waves are observed in the hypothalamus.

4. Polarization of Individual Layers of the Cortex

The role of different layers of the cortex in the creation of a dominant focus by dc polarization can be investigated for the motor cortex with a current of minimal strength (0.5 μA), Ryabinina (1961) demonstrated that various layers have different roles in the formation of a motor dominant focus. Her experiments were carried out on adult rabbits under chronic conditions. Having confirmed that surface, anodal polarization produces a focus of excitation with the properties of a dominant, she showed that during polarization of the lower layers of the motor cortex the clearest results were obtained if the tip of the polarizing electrode was in layer V. Under these circumstances the dominant focus appeared only for cathodal polarization. The criterion of formation of the dominant focus, as in the other experiments, was the appearance of motor responses of the forelimb on the contralateral side relative to the focus, on application of acoustic or photic stimuli to which the orienting reflex had first been extinguished. Polarization with the electrode in layer VI, or in the white matter, either gave no effect, or the effect was extremely tenuous. If the electrode was in layer V, the dominant properties appeared very rapidly after the beginning of polarization. Prolonged action of the cathode on the pyramidal cells led to inhibition of the dominant focus: the rabbit ceased to respond to acoustic and photic stimulation. After the end of the cathodal polarization a "rebound" effect was observed: the rabbit responded to hitherto indifferent stimuli by an increased motor response evoked. In these experiments with cathodal polarization of the pyramidal cells the following changes took place in the EEG of the superficial and deep levels of the motor cortex: spindles disappeared or were reduced in size; the amplitude was reduced; during the action of indifferent stimuli synchronized rhythms appeared in the EEG. Synchronization at the cortical level during a motor response in the presence of a dominant focus was also observed by Meshcherskii (1951), and during conditioning by Knipst (1955).

With anodal polarization of deeper layers of the motor cortex as a rule no dominant focus appeared. Usually high-amplitude irregular waves appeared in the EEG. The action of the anode directly on the pyramidal cells of layer V inhibits foci of excitation previously formed there by weak cathodal current, and this inhibition continues for up to 30 min.

These experiments show that the lower layers of the cortex, and specifically layer V of the motor cortex, play the most important role in the formation of the dominant focus during dc polarization. Histological tests showed that when the polarizing electrode was in the layer of large pyramidal cells the clearest evidence of a dominant focus was obtained. In any other position, either no dominant focus was formed whatever, or only weak signs of increased excitability were observed in the motor area, as shown by the weak and irregular motor responses of the forelimb to indifferent stimuli. In this respect the results agree with those obtained by others who demonstrated the excitatory effect of cathodal polarization on cortical neurons of the lower (V-VI) layers of the cortex (Burns, 1954; Beritov and Roitbak, 1954).

Some interesting investigations were carried out by Landgren et al. (1962a, b) and Hern et al. (1962) to study differences in anodal and cathodal stimulation of the cortical surface in monkeys. Like Fritsch and Hitzig (1870), they were struck by the fact that the anode stimulates when applied to the cortical surface. Phillips and his colleagues used currents stronger than those discussed above, e.g., 1-5 mA versus 1-10 μA. The most striking difference, according to Phillips' observations, between the anode and cathode on the cortical surface is seen during repetitive stimulation by rectangular pulses up to 5 msec in duration. The threshold with the cathode on the cortical surface is higher than for the anode. Repetitive anodal stimulation, as intracellular recordings from motoneurons showed, evokes a series of equal synaptic potentials. After the end of stimulation there is no aftereffect. A different response is obtained to repetitive cathodal stimulation. Synaptic potentials increase in amplitude and display discontinuity in both the ascending and descending phases. On removal of the stimulus no spikes are obtained from the motoneurons but its membrane continues to receive a subthreshold synaptic bombardment of the corticofugal afterdischarge.

For anodal stimulation the responses owe their origin evidently to an outward current through the excitable membrane in the region of the pacemaker of the pyramidal cell. Similar results, obtained on receptor organs (Maruhashi, Mizuguchi, and Tasaki, 1962), also are explained by stimulation of an outward current through the axon close to the receptor organ. It may be, as Phillips suggests, that the anode on the cortical surface hyperpolarizes the apical dendrites and depolarizes the axons of the same cells. If a cathode is applied to the cortical surface, systems of axons and neurons in the surface layers of the cortex are excited. Cathodal stimuli evoke responses with a longer latent period and they are more variable, evidently because of indirect excitation of the pyramidal cells. The optimal focus for the cathode is always in the precentral direction relative to the optimal focus for the anode. If an anode, in its optimal focus, excites pyramidal neurons, a cathode in the same place inhibits them.

The afterdischarge obtained with a cathode is connected with activation of the intracortical mechanism. If the appearance of this phenomenon were due to an interneuronal circular connection at the spinal level, this intraspinal reverberation would appear irrespective of the polarity of the cortical stimulation which evokes the pyramidal discharges. From experiments in which action potentials were recorded from single pyramidal tract fibers and synaptic potentials of spinal motoneurons, Phillips et al. concluded that the anode directly excites the spike generator or the pacemaker region of the pyramidal cells, while the cathode excites the pyramidal cells through synaptic mechanisms. These experiments confirmed the presence of monosynaptic pyramidal pathways in primates and showed that pyramidal neurons are in fact connected monosynaptically with α-motoneurons in the spinal cord. These cortico-motoneuronal pathways are able to fractionate the segmental reflex into small, isolated movements along the lines discussed by Leyton and Sherrington (1917), and they can also initiate or, in emergency, inhibit a movement.

The importance of the polarization mechanism in the formation of a flexor dominant focus was demonstrated by Verzilova (1966).

5. The Recruiting Response

Stimulation of the nonspecific thalamic system by rectangular pulses (0.1 sec, 4-6 V) evokes a recruiting response, which is recorded mainly from the cortical surface in the sensorimotor area. This response has been well studied and described in the literature. Under optimal conditions recruiting responses are characterized by a progressively increasing surface-negative wave, reaching its maximum after 2-5 successive stimuli. If stimulation continues the responses may decrease progressively in amplitude, and then recover again, thus resembling in their form "spindling" activity. Recruiting responses are not always surface-negative. They may also be biphasic, with an earlier positive phase of low amplitude. The early positive phase may depend on simultaneous stimulation of certain specific projection fibers which evoke the initial surface-positive responses with a shorter latent period. The latency of the surface-negative wave of the recruiting response is usually 20-40 msec after stimulation of the thalamus.

Recruiting responses recorded from the cortical surface are regarded by most authorities as analogs of "synaptic" or "dendritic" potentials, for they are independent of discharge of cortical neurons recorded by microelectrodes from single units. The initial waves of the recruiting response are often recorded before the unit discharge. This indicates that these waves are primary, responsible for the unit discharge, and not vice versa.

As already mentioned, the "nonspecific" thalamic projection system in fact has a very definite and precise, "specific" organization of its cells and fibers. With stimulation of the medial thalamic nuclei by weak currents, sometimes moving the electrode a fraction of a millimeter is sufficient to obtain a different response when recording from the cortical surface.

Results obtained by Pavlygina showed that, if the place of stimulation in the medial thalamic nuclei of the cat remained constant, the optima of strength and frequency of stimulation differed for different points of the motor cortex (Fig. 27). If the place of recording the recruiting response remained constant (recording

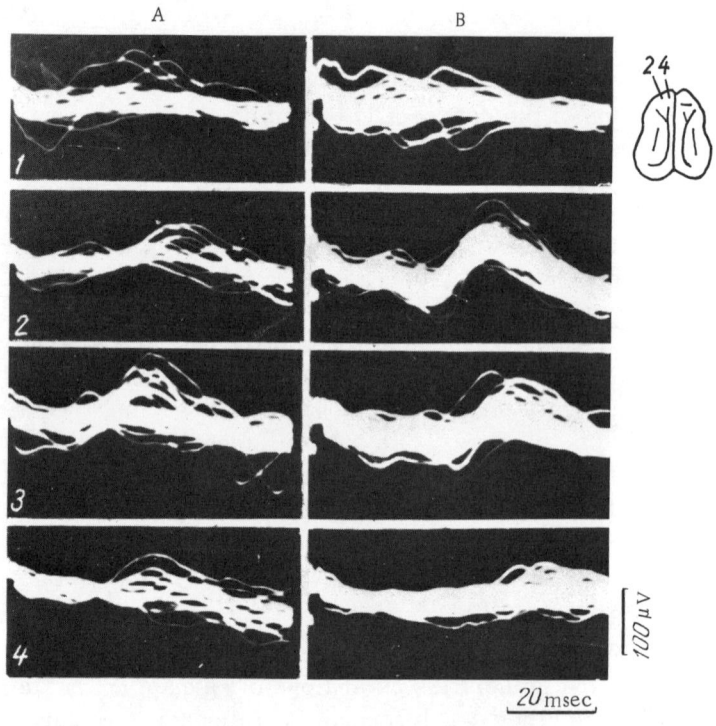

Fig. 27. Recruiting response in different points of motor cortex in the cat (work of Pavlygina): A) medial thalamic nucleus stimulated by 3V, 0.1 msec, successively at 4 (1), 6 (2), 8 (3), and 19 Hz (4). Recording taken at point 2 (diagram), superposition of 25 responses. Optimal frequency of stimulation 8 Hz. B) medial thalamic nucleus stimulated by 5.6V, 0.1 msec. Recording from point 4. Optimal frequency of stimulation 6 Hz.

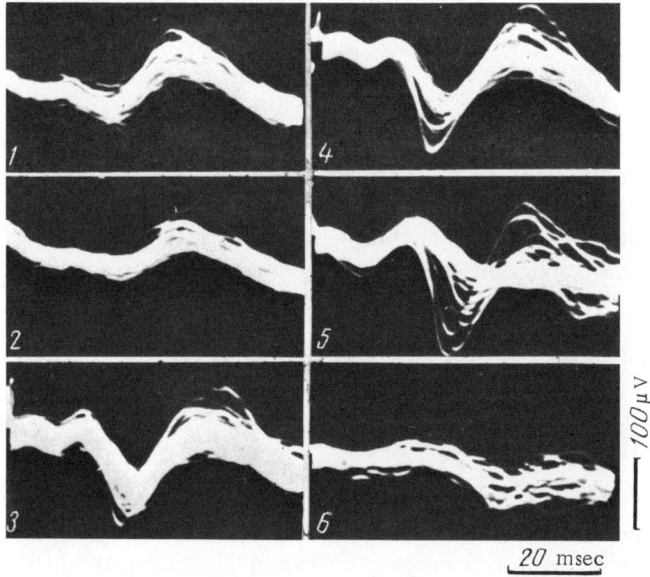

Fig. 28. Changes in shape of recruiting responses recorded at same point of cat's motor cortex ("monopolar" recording with "bipolar" stimulation (5.6 V, 0.1 msec, 6 Hz) at different levels of the medial thalamus (work of Pavlygina): 1-6) stimulating electrodes moved from above down in consecutive steps of 1 mm. Superposition of 25 responses.

electrode on the cortical surface at the place where the response was recorded best), minimal vertical movement of the stimulating electrodes in the medial nuclei of the thalamus altered the recruiting response (Fig. 28). These results indicate structural and functional differentiation of the nonspecific thalamo-cortical system.

At the International Physiological Congress in Leiden (1962), Caspers (1962) reported that during an investigation of changes in cortical steady potential he had observed slight differences in the functional organization of the reticular system which were not detectable on the ordinary EEG. He showed that during stimulation of the reticular formation with high-frequency pulses, the cortical surface becomes more negative, and that the process has a steep rising phase at the beginning of stimulation and only very slightly outlasts the stimulation. A response of this type is obtained mainly for stimulation of the medial divisions of the reticular formation, which according to Brodal (1960) are characterized by neurons with long axons. However, at the same time, the cortical surface gradually begins to become negative after a latency of several seconds. The amplitude of this type of negative shift reaches a maximum a few seconds after removal of the stimulus. This type of response is predominant for stimulation of more lateral parts of the reticular formation, where neurons with short axons are to be found. They rapidly become adapted to repetition of the stimulus.

In Pavlygina's experiments evidence was obtained of some differences in the functional organization of the diffuse system of the thalamus which cannot be detected in the ordinary EEG, but which are found from the recruiting response as the electrodes are moved vertically.

To determine the role of the diffuse system of the thalamus in the formation of the dominant focus, when a focus of excitation directs the organism's response toward a particular reflex regardless of the position or nature of the stimulus, it is important to know the principles governing interaction between the "nonspecific" and "specific" thalamo-cortical systems. This problem has been solved by investigating the changes in visual evoked responses and recruiting responses during simultaneous application of flashes and stimulation of the medial thalamus. There is evidence in the literature that such a combination of stimuli evokes a consistent response: stimulation of the medial thalamus evoking a recruiting response facilitates evoked responses to light.

If the question of a consistent response in the nervous system is raised, it is always followed by the further questions: does the response exhibit an altered pattern if the conditions of stimulation are changed; is there an opposite response if the state of the stimulated structure is changed? Pavlygina's experiments showed that there is no consistent response as regards the amplitude of the evoked responses to photic stimulation during recruiting. The following principle is obtained: if an evoked response to photic stimulation recorded in the occipital region is of high amplitude, medial thalamic stimulation will not increase but, rather, will reduce it. In other words, thalamic stimulation evoking a recruiting response, may, to judge from the effect it has on evoked cortical potentials, have both a "facilitatory" and an inhibitory action on the

5. THE RECRUITING RESPONSE

specific afferent system. The effect depends on the functional state, background, and level of excitability of the structure stimulated.

Simultaneous stimulation of the medial thalamus and application of flashes has a dual action: not only the photically evoked potentials are modified, but also the recruiting response itself. Photically evoked potentials recorded from the occipital cortex in cats, are modified in the following manner by thalamic stimuli evoking a recruiting response in the motor cortex. The negative phase and the subsequent negative wave are increased, and the latent period of the subsequent negative wave is lengthened. The recruiting response is modified by simultaneous thalamic stimulation and application of flashes of the same frequency in such a way that each response is followed by a negative wave. The clearly defined interaction of photic stimulation and stimulation of the medial thalamic nuclei, when applied simultaneously, is noteworthy. The technique of stimulation is such that the beginning of the pulse stimulating the thalamus triggers both the flash and the sweep of the oscilloscope (Fig. 29). Interaction between the two thalamic systems when stimulated simultaneously evidently indicates the electrotonic nature of this phenomenon.

Another phenomenon, which occurs regularly after repeated combinations of stimulation of the medial thalamic nuclei and simultaneous flashes, must also be considered. If the combinations are stopped and flashes only are applied, a large negative slow

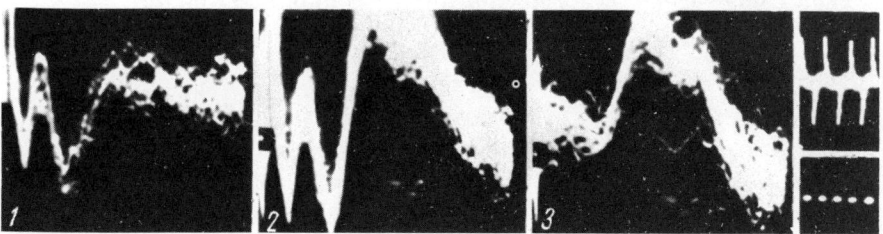

Fig. 29. Changes in recruiting responses by concurrently presented flashes (cat; work of Pavlygina). 1) recruiting response without flashes; 2) recruiting response plus flashes; 3) response of motor cortex to flashes after repeated combinations of flashes and stimulation of the medial thalamus. Superposition of 25-30 responses. Calibration: 20 msec, 100 µA.

wave appears in the motor area at each flash (Fig. 29). Evoked potentials to a photic stimulus are also recorded in the motor cortex without preliminary combination of the photic stimulus and stimulation of the medial thalamus. These are secondary potentials spreading through the diffuse nonspecific system. However, in these experiments they did not appear to every photic stimulus and were not so clearly marked.

These results are evidence that the nonspecific thalamo-cortical system plays a role in the formation of temporary connections, for the considerable increase in amplitude and stabilization of the secondary evoked potential appearing in the motor cortex of cats in response to combinations of two stimuli, one of which evokes potentials in the motor cortex, the other in the occipital cortex, is an indisputable fact. To express it more precisely, these findings are regarded as the result of interaction between two thalamo-cortical systems – specific and nonspecific – interaction which is electrotonic.

6. Functional Connections of the Non-Specific Thalamo-Cortical System with Cortical Neurons

In the course of this work the question arose of functional connections between the nonspecific thalamo-cortical system and cortical neurons, and also of the connections of the extrapyramidal system with the cortex and reticular formation and the role of these connections in the formation of the dominant focus.

In 1938 Lorente de Nó described three principal types of afferent fibers in the parietal cortex of adult mice. Shkol'nik-Yarros (1965) investigated endings of afferent fibers in the cortex in more detail and gave a fuller classification. According to Lorente de Nó these are: association fibers, specific, and nonspecific or diffuse fibers. In their course in the white matter the nonspecific fibers give off collaterals, and as they enter the cortex give off branches at all levels. Their endings can be traced without interruption as far as the molecular layer. When Dempsey and Morison (1942) described cortical recruiting responses to rhythmic stimulation of the medial thalamus in the cat, attention was drawn to these almost forgotten observations of Lorente de Nó and the presence of a nonspecific thalamocortical system was postulated, although no connections were demonstrated between the intralaminar nuclei of the thalamus and fibers of this type in the cerebral cortex.

The extensive influence of regions of the diencephalon on cortical activity has been known ever since the work of Hess (1929, 1944, 1949), Dempsey and Morison (1942), Murphy and Gellhorn (1945), and Jasper et al. (1948). These investigations showed the effect of electrical stimulation of the intralaminar nuclei and hypothalamus on cortical electrical activity.

Some authors regard the diencephalon, especially the nonspecific thalamic nuclei, as the "higher" sector of the reticular system. Moruzzi and other Italian investigators, who are responsible for much of the work on the reticular system, consider that this view is supported by neither anatomical nor physiological findings. The diffuse thalamic projection system is only one of the structures which receive ascending reticular fibers, as the hypothalamus also does. Long ascending reticular fibers also run directly to the corpus striatum, bypassing the diencephalon.

Destruction of both hypothalamus and thalamus does not prevent the onset of an arousal reaction to stimulation of the reticular formation. These workers thus maintain that the concept of an "activating system," including structures in both the brain stem and diencephalon, cannot be accepted, at least at the present level of knowledge. From the evolutionary point of view, on the basis of Kurepina's (1959) findings, there is no valid reason for regarding the "thalamic reticular system" as the direct continuation of the mesencephalic reticular formation, for they are of different origin.

Jasper applied the term "reticular formation of the thalamus" to its paraventricular substance, intralaminar cell concentrations, and the reticular zone.

Since Demsey and Morison (1942) described "diffuse" cortical responses evoked by stimulation of certain parts of the thalamus, many investigations, both physiological and anatomical, of the so-called nonspecific system of the thalamus have been published. Jasper (1949) found that stimulation of the thalamus at the "critical point" evokes bilateral responses with complete control over spontaneous cortical rhythms. When, however, the electrodes are placed 3-4 mm laterally to the midline, bursts of activity are recorded predominantly in the ipsilateral hemisphere, and with a longer latency contralaterally. If the anterior pole of the reticular nucleus is stimulated, bursts appear in one hemisphere or in a more restricted part of one hemisphere. The predominantly ipsilateral or bilateral effect on the cortex can be evoked by different parts of the same system. It is interesting to note that Dempsey and Morison (1942), although mentioning the diffuse character of the recruiting response, make no mention of the bilateral effect. They clearly established the fact that recruiting potentials appear in all regions of the ipsilateral hemisphere. Hanbery and Jasper (1953) found that stimulation of the nonspecific thalamic nuclei evokes a widespread response, involving nearly the whole of the ipsilateral cortex and anterior parts of the contralateral hemisphere. No pathways to the anterior cortical regions of the contralateral hemisphere have yet been discovered. There is relatively little information which can be used as evidence for a bilateral recruiting response. Although it is well recognized that repetitive stimulation of the nonspecific system of the thalamus has a mainly diffuse and bilateral effect, this system does not behave uniformly throughout its extent. As Jasper (1949) pointed out, different results are obtained depending on which part of this system is stimulated.

In practice it is relatively easy to evoke a typical ipsilateral recruiting response, whereas true diffuse and bilateral recruiting is evoked much less commonly.

Enomoto (1959), working with cats, determined from which parts of the nonspecific thalamic system a bilateral effect can be evoked by repetitive stimulation. Stimulation of those lying closer to the midline evokes a bilateral effect much more commonly, whereas stimulation of other nuclei, situated more laterally, rarely evokes a contralateral effect. Like others, Enomoto found that the pathways for the bilateral effect run mainly, but not entirely, through the anterior part of the "massa media."

On the basis of published work and of his own experiments, Durinyan (1964) introduced several modifications into the existing classification of the thalamocortical projection systems, taking account of their functional properties. These changes apply mainly to the nonspecific system, for the diffuseness and nonspecificity of

6. FUNCTIONAL CONNECTIONS

this system are unconvincing. Durinyan removes the thalamic reticular formation from the nonspecific system and regards the nonspecific system itself as secondary and associative in character.

Despite the extensive literature, little is yet known on the functional relations between the nonspecific thalamic nuclei and cortical neurons. Li (1956) found that stimulation of the nonspecific thalamic nuclei evokes "facilitation" of cortical single unit responses. Some conclusions regarding anatomical connections of the nonspecific fibers with cortical neurons were also drawn from his investigations. Nonspecific thalamic impulses activate cortical "interneurons," which do not respond to stimulation of the thalamic sensory relays. According to Li's (1956) findings, nonspecific thalamic impulses exert their effect on the activity of cortical sensory neurons through these interneurons. In his later investigations, Li (1962) demonstrated essentially the same phenomenon for neurons of the motor cortex. Nonspecific thalamic nuclei activate pyramidal neurons via interneurons, which do not themselves send axons into the pyramidal tract but increase the excitability of the pyramidal neurons.

Activity of the extrapyramidal system is also a factor relevant to cortical-subcortical relationships in the presence of a dominant focus. Interaction between cortex and basal ganglia in the formation of a motor dominant is of practical importance, for it helps to explain the various cases of hyperkinesia in patients with lesions of the striopallidal system. However, so far, the influence of the extrapyramidal system on the cortex, and the anatomical and functional connections between the various nuclei of the extrapyramidal system cells and with the nonspecific system, still remain largely unexplained. The available literature is conflicting. The large number of fibers passing through the basal ganglia (caudate nuclei, putamen, globus pallidus, etc.) or near them makes precise evaluation difficult, because destruction of one part leads to simultaneous injury to fibers running close by.

Using the technique of neuronography Ryabinina (1963) shed some light on the connection between the extrapyramidal system and cortex. Application of strychnine to the rabbit's motor cortex evokes strychnine spikes both in the region of application and also in other structures with direct anatomical connections with the strychninized cortex. The same electrodes, consisting of glass cannulas 50μ in diameter, were used to record the potentials and to inject microdoses of strychnine.

Injection of strychnine into the globus pallidus evokes high-voltage discharges not only in the globus pallidus itself, but also in the mesencephalic reticular formation and motor cortex. Injection of strychnine into the mesencephalic reticular formation evoked discharges in the globus pallidus and in the motor cortex. It thus follows that functional connections exist between the globus pallidus and cortex and also between the mesencephalic reticular formation and the cortex. The globus pallidus and mesencephalic reticular formation are thus included in the functional organization of the motor system.

A focus of excitation produced by strychnine in the mesencephalic reticular formation can become a dominant focus. It begins to summate excitation from hitherto indifferent stimuli, after which the arc of a motor reflex can be closed by connection between the mesencephalic reticular formation and the extrapyramidal system, so that the animal begins to give a motor response to such stimuli.

These results shed some light on the physiological mechanisms of hyperkinesia, a phenomenon which closely resembles the pathological dominant focus.

* * *

Like polarization of the cortex and other subcortical structures, polarization of both the nonspecific and specific thalamic nuclei can produce excitation or depression depending on its strength. Interaction between these two thalamic systems when they are stimulated exactly simultaneously is thus apparently electrotonic. The electrotonic character of interaction in the central nervous system is also demonstrated by the appearance of bilateral slow potentials in the motor and visual areas of the cortex during defensive conditioning to photic stimulation (see Chapter II). Besides cortico-striatal connections there are also pallido-cortical connections, which are responsible for one of the feedback mechanisms in cortico-subcortical interaction. This "loop" may be the source for the formation of a focus of stationary excitation. When connected with the "circuit" of brain stem reticular formation —

globus pallidus — brain stem reticular system, or with the longer "circuit" traveling through the diffuse system of the thalamus and other basal ganglia, this central focus of excitation, and the pathological condition, may form the "mechanism" of hyperkinesia.

Chapter VI

Relations between Specific and Nonspecific Thalamo-Cortical Systems. Correlation Analysis of EEG Rhythms

1. Relations Between the Specific and Nonspecific Thalamo-Cortical Systems in Man Based on the Study of Evoked Potentials

Foci of excitation in the central nervous system which become dominant foci, hence yielding the same reflex response regardless of the place and type of stimulation, present a conceptual problem since a diffuse spread of excitation is present concurrently with a local excitatory condition. This, in turn, raises the question of the dynamic relations between the specific and nonspecific afferent systems.

It is claimed in the literature that the nonspecific projection system is independent of the specific thalamic nuclei. The evidence for this claim is that a recruiting response can be recorded from each region of the cortex, including the sensory areas, after complete destruction of the thalamic nuclei with specific connections with the individual cortical areas. These results and many others led some investigators to recognize competitive relationships between the two thalamic systems.

The relationships between the specific and nonspecific thalamo-cortical systems have been investigated in man by recording evoked potentials. The opportunities for studying the healthy human EEG are limited to gross recordings from the surface of the head. Subcortical, brain-stem, and other deep brain structures are, as a rule, inaccessible to investigation. Only in rare pathological cases can deep recordings be obtained from individual subcortical structures during neurosurgical operations. Nevertheless, several aspects of relations between specific and nonspecific thalamocortical systems can be solved by electroencephalography in man; they can in fact be examined in a more precise form than in animals.

When evoked potentials were recorded from the surface of the human head, the following facts were observed. Besides potentials evoked in response to stimulation of one modality, nonspecific potentials also appeared in response to stimulation of a different modality. Nonspecific responses to acoustic, photic, and other stimuli occur mainly in the central-parietal region. Responses of this type were first described by Davis (1939) and later by Gastaut (1953), Bancaud et al. (1953), and others. The view that the nonspecific potential is produced by displacement of the skin relative to the skull by movement of the ears (Oswald, 1959) is disputed by Larsson (1960) and Kats (1958a). The nonspecific response reflects the activity of the nonspecific thalamo-cortical system. By investigating the relation between the nonspecific and specific response in the human EEG, the relations between these two thalamic systems can be judged.

Kats, working in this laboratory, showed that the nonspecific response is greatest in the parieto-central region, but with repetition of the stimulus it spreads to other regions. This spread takes place simultaneously with an increase in the amplitude of the response (Rusinov, 1962). In other words, Kats showed that the nonspecific response to the EEG is not strictly localized, as was hitherto considered. Puchinskaya (1964) investigated nonspecific responses in the human EEG and showed that during repeated flashes the nonspecific response disappeared in the zone where it usually occurred, while in the occipital region it was recorded more clearly, i.e., the nonspecific response to photic stimulation was displaced into the posterior cortex. For the first flashes the nonspecific response was diffuse, and its amplitude was maximal in the central region. For subsequent flashes, the nonspecific response shifted posteriorly. The suggestion was made that this shift of the nonspecific response is associated with increased excitability in the posterior parts of the hemisphere under the influence of the photic stimulation itself. This hypothesis was confirmed by other evidence. If an acoustic stimulus was presented against the background of repetitive photic stimulation, the nonspecific response appearing to this acoustic stimulus also shifted posteriorly, i.e., into the projection region of the repetitive photic stimulation. With repetition of the acoustic stimulus alone, without photic stimulation, no such phenomenon occurred.

The specific response to a flash is recorded in the occipital regions, the nonspecific response mainly in the anterior region. In cases of brain pathology, when an irritative focus (such as an extracerebral tumor) is present, the nonspecific response is recorded clearly in the zone of the focus. These facts can be explained as follows. Impulses evoking the nonspecific response affect the cortex diffusely. The response arises, however, in the region of increased excitability. In the case of repeated flashes or any repetitive photic stimulation, a focus of increased excitability is formed in the occipital region, and this is reflected in the appearance of a nonspecific response in that region. A lesion, especially a localized extracerebral tumor, can create a localized state of increased excitability. This evidently accounts for the appearance of the nonspecific response in the region of the lesion. This state is analogous to that obtained in the presence of a cortical dominant focus, when a stimulus of any modality evokes a nonspecific response in that same region of the cortex.

Foci of excitation assuming the character of dominant foci are observed in several pathological conditions. Grindel' and Filippycheva (1959) investigated disturbances of human motor activity in patients with tumors in the frontal lobes and demonstrated stationary excitation, resembling a dominant focus, in the affected motor cortex. They showed that flashes presented against the background of a rhythmic motor response augment the movements. The same summation was also obtained by the use of other weak stimuli, tactile for example, thus demonstrating that the pathological focus behaves as a dominant focus.

The facts obtained by Kats and Puchinskaya, in this laboratory, were subsequently confirmed. Under various experimental conditions Raeva (1966) investigated the responses evoked in the human

EEG by repeated tactile or acoustic stimuli, or combinations of both. When analyzing her results she concentrated mainly on the intensity of the responses in different cortical regions. As tactile stimulus she used a pneumatic vibrator fixed to the subject's right hand over the thenar eminence. The CS was a combined acoustic stimulus consisting of three stimuli each with a frequency of 500 Hz at 58 dB, presented at equal intervals. In response to the tactile stimulus, a well-defined potential corresponding in its parameters to the nonspecific response described in the literature was recorded in the EEG in the postcentral region of both hemispheres. In response to the first presentation of the tactile stimulus, the potential evoked was usually bilateral, and equal in the postcentral region of both hemispheres. Later, on repeated application of the tactile stimulus, the evoked potential moved and was redistributed: gradually becoming unilateral, and was recorded only in the hemisphere contralateral to the stimulus, i.e., in the cortical projection zone of the right hand.

These observed local changes were reproduced by a CR combination of a previously extinguished, indifferent acoustic stimulus with tactile reinforcement of the right hand. During the first combination the acoustic stimulus caused no appreciable change in the EEG. Tactile stimulation, however, evoked a well marked potential in both hemispheres, followed usually by generalized depression of the original rhythm. In the course of subsequent combinations, similar local changes in the EEG began to appear by a CR mechanism in response to the acoustic stimulus before application of the tactile stimulus. Later, the facts described previously were observed: a gradual migration, with an increase in the number of combinations, of the zone of greatest intensity of responses to the contralateral hemisphere, into the cortical projection zone of the US.

Thus, besides depression of the alpha-rhythm, evoked potentials of the nonspecific response type are recorded in the human EEG in response to tactile stimulation. These potentials change during repeated stimulation: initially they are recorded bilaterally, in the postcentral regions of both hemispheres, but later they are recorded predominantly in the contralateral hemisphere. Similar changes in the potentials, with migration to the contralateral hemisphere, are recorded as the result of a CR.

These phenomena can be explained by the hypothesis of foci of excitation in the central nervous system. A whole series of factors was evidently concerned in the formation of the "focus" in the corresponding structures of the human sensorimotor cortex: repetitiveness of the stimuli, their specificity, and the functional state of the central nervous system. The results suggest, like Puchinskaya's findings, that the direction of impulses from the nonspecific thalamic system under certain conditions may become selective, and may be determined by the state of excitability of the corresponding cortical projection structures.

Sokolova (1965), also working in this laboratory, using scalp electrodes investigated potentials evoked by electrodermal stimulation in healthy subjects and in patients with local brain lesions. Stimuli were applied to the skin of the hand by means of a Multistim electronic stimulator. Evoked potentials were recorded in the healthy subjects as biphasic or triphasic waves with a total duration of 200-500 msec. They were most marked in the central and frontal leads. As a rule they occurred bilaterally. Comparison with data in the literature (Allison et al., 1962) shows that these bilateral potentials are the nonspecific response to electrodermal stimulation. They also appeared in response to stimulation of other modalities (photic, acoustic).

In patients with a local brain lesion, and with marked EEG responses to afferent stimulation, definite changes in the responses were found in every case by comparison with the responses recorded in healthy subjects: asymmetry of the evoked potentials in the two hemispheres was observed, the responses being greater on the side of the pathological focus. In this case stimulation was applied deliberately to the ipsilateral hand relative to the pathological focus so as to exclude any possible predominance of evoked responses in the hemisphere contralateral to the side of stimulation. Investigation of potentials evoked by photic and acoustic stimuli in the same subjects showed that these stimuli also yielded potentials of unequal intensity in the two hemispheres, their amplitude being higher on the side of the pathological focus.

If a focus of pathological electrical activity is present in one occipital region, asymmetry relative to evoked potentials is seen most clearly in the occipital regions, with marked predominance in the hemisphere on the side of the lesion. There, in the occipital region on the side of the lesion, the evoked responses both to electrodermal, and to photic and acoustic stimulation are well marked.

* * *

The potentials evoked to photic, acoustic, tactile, and electrodermal stimulation in healthy subjects and patients with organic brain lesions thus indicate that the ultimate effect of an afferent stimulus evoking a nonspecific response is decided in the cortex, and that the mechanism of this decision is that of a focus of excitation which acquires the properties of a dominant focus. The nonspecific evoked response shifts to the side of the focus of excitation, indicating selective manifestation of the nonspecific potential in the cortex. The fact that the nonspecific response shifts in the direction of the primary projection area of the specific system during repetition of the stimulus does not confirm the view that the two thalamic systems are competitive, but is evidence of their concerted, coordinated activity and of the dynamic relationship between them.

2. Correlation Analysis of the EEG of the Motor and Visual Cortex during Polarization of the Motor Cortex

Events taking place in the motor cortex when a focus of excitation changes into a dominant focus characterize only one aspect of the system formed in the course of the response. The system also includes the specific cortical projection area of the afferent stimuli reinforcing the focus.

The work of Kalinin (1965) showed that during the formation of the response of a motor dominant focus to photic impulses, secondary evoked potentials in the occipital region are inhibited. The dominant focus in the motor cortex is reinforced not only by impulses of different modalities, traveling through the reticular system directly or via interneurons to the pyramidal cells, but also with the participation of the primary specific cortical projection area of the reinforcing stimuli.

The question of correlation between the motor and visual cortex during polarization of the motor cortex must naturally be considered. There is a solid basis for correlation analysis in electroencephalography in both the Soviet and Western literature (Wiener, 1961; Barlow, 1961; Brazier and Casby, 1952; Kozhevnikov and Meshcherskii, 1963; Sokolov, 1965; Danilova, 1964; Mishin, 1963; Grindel' et al., 1964, 1965, 1966). Such investigation of the EEG involves the study of quasi-periodic processes (the basic, dominant activity), true periodic processes (cortical evoked responses to repetitive photic stimulation), and aperiodic variable processes. These are the same classes of temporal functions as are encountered in information theory. Of the many theories in modern information science, that most suited to investigation of the EEG, as Wiener has shown, is the theory of auto- and cross-correlation functions.

The changes in the degree of communication between the cortical motor and visual areas of the rabbit during polarization of the motor cortex were investigated by EEG correlation analysis in this laboratory by Zhegalkina (1967). Chronic experiments were carried out on animals with electrodes implanted on the dura. The EEGs were analyzed by a correlometer designed by the All-Union Research Institute of Medical Instrumentation (VNIIMIIO). From the results of these experiments it was possible to assess the predominant frequencies, compare the processes investigated as regards their degree of periodicity, and to study the degree of connection between regions. The relative increase in the coefficient of cross-correlation was also used as an index of the degree of change in the level of communication between different cortical regions.

The changes in the spectral composition of the EEG during polarization is clearly seen in autocorrelograms. The low-frequency rhythm of the order of 5 Hz, predominant before polarization (2 μA) in both the visual and motor areas is sharply reduced during the first minute of polarization, while rhythms of higher frequency (12-13 Hz in the motor cortex, 10 Hz in the visual cortex) appear, together with a weak rhythm of the order of 3 Hz. Later in the course of polarization, the activity of the high-frequency waves is reduced and the low-frequency rhythm of about 3 Hz becomes stronger (Fig. 30).

An interesting feature in these experiments is the maximum absolute value of the cross-correlation coefficient (C_{cr}), which is an index of the degree of connection between the motor and visual

156 CHAPTER VI: SPECIFIC AND NONSPECIFIC THALAMO-CORTICAL SYSTEMS

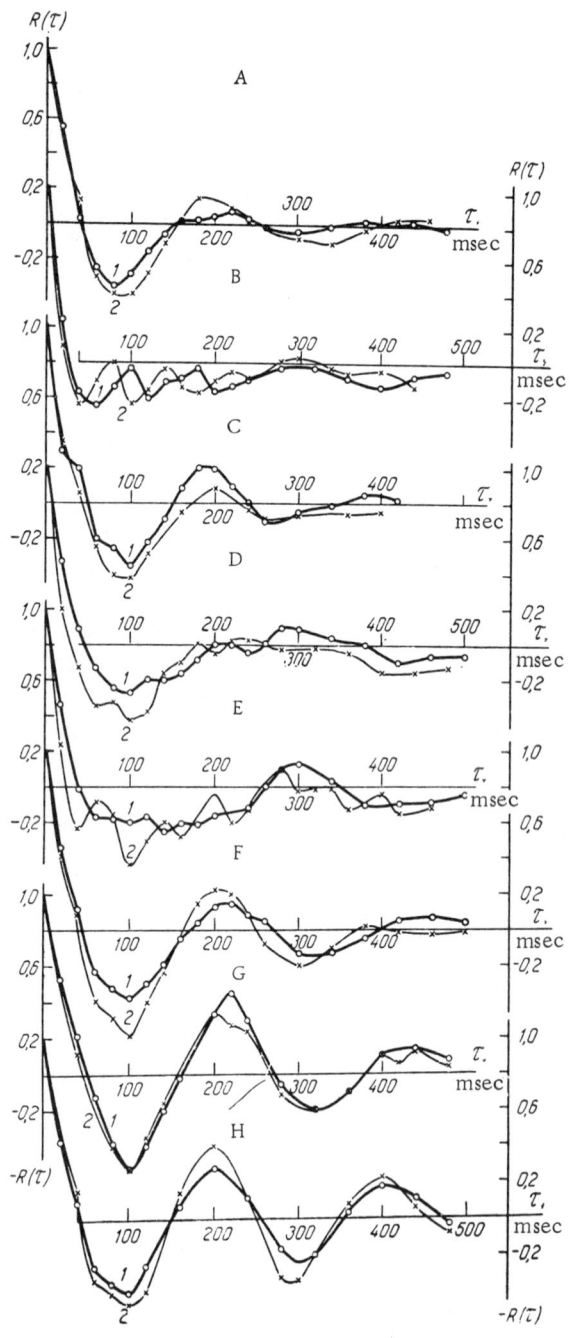

Fig. 30. Autocorrelation functions of the EEG recorded from the visual and motor cortex of a rabbit: A) before polarization; B) 1 min, C) 10 min, D) 20 min after beginning of polarization of motor cortex by direct current of 2 μA; E) 1 min, F) 15 min, G) 30 min, and H) 60 min after removal of direct current. 1) visual cortex, 2) motor cortex.

areas. Comparison of the absolute maximum value of the C_{cr} of the EEGs of the motor and visual areas before and after polarization (2 μA) reveals a definite relationship between the initial level of connections and the changes affecting the C_{cr} at onset of polarization of the motor cortex. In other words, a relationship was established between C_{cr} of the background (before polarization) and the difference C_{cr}(polarization) − C_{cr}(background). If C_{cr} before polarization is less than 0.6-0.7, the cross-correlation coefficient rises; if C_{cr} before polarization is greater than 0.6-0.7, the oppo-

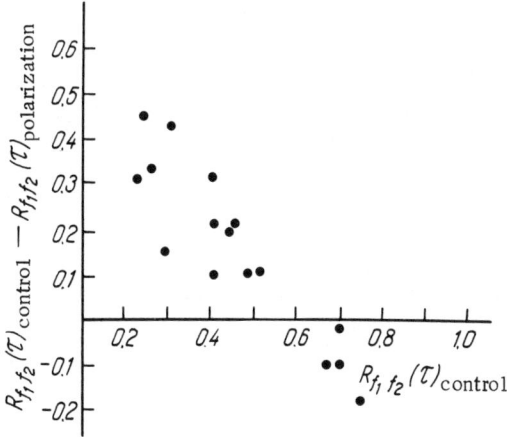

Fig. 31. Dependence of degree of change in coefficient of cross-correlation between EEGs of sensorimotor and visual areas during polarization (2 μA) on initial level of connection between between these areas. First minute after application of polarizing current (sensorimotor area) (work of Zhegalkina).

site is observed: the C_{cr} decreases, i.e., the sign of the increase in C_{cr} is reversed (Fig. 31).

These facts suggest that an optimal level of connection corresponding to specified polarization conditions exists between the cortical projection zones. This hypothesis is in good agreement with the results obtained by other methods (Rusinov, 1965).

3. Foci of Increased Excitability Evoked in the Cortex by Repetitive Stimulation

Foci of increased excitability in the cortex, evoking phenomena in which "a particular focus in the cortex loses its role as an apparatus with a single function" (Ukhtomskii, 1911), were investigated electrographically by Roitbak (1955). During tetanization of the cortical surface (in cats lightly anesthetized with nembutal) rhythmic electrical activity arising in a localized area of cortex is observed. The same stereotyped response can be seen over and over again for dozens of times. During tetanization of the cortex the rhythmic electrical activity does not spread further than a few millimeters, but after the tetanization is discontinued, the activity spreads beyond this limit. Evidently during tetanic stimulation, evoking and maintaining a focus of rhythmic activity, an inhibitory process of "active localization" took place.

In his next experiments, Roitbak (1953, 1955) combined stimulation of two points of the cortex. A pair of stimulating electrodes was applied to the posterior pole of the suprasylvian gyrus and another pair to the anterior pole of the same gyrus, near the sensorimotor area (the first and second pairs respectively). The recording electrodes were halfway between them. During stimulation through the first pair at 9 Hz, waves of the same rhythm were recorded. Without interrupting the first stimulation, tetanic stimulation was then applied through the second pair of electrodes at 50 Hz, and initially each stimulus evoked a negative wave. For the first few seconds of tetanization, inhibition developed around the stimulated point of the cortex: "spontaneous" electrical activity was inhibited and the effect of stimulation of other regions of the cortex was weakened or disappeared. After a few seconds of tetanization, its direct effect disappeared but strong rhythmic activity arose at 9 Hz. This activity differed from the usual effect of stimulation through the same pair of electrodes in its longer latent period and higher amplitude. The direct effects of stimulation (the first pair) which were produced before the stimuli were combined still continued to appear, but their amplitude was lowered. As a result of tetanic stimulation (the second pair), excitability must be greatly increased in a certain group of cortical neurons, in which a focus of rhythmic electrical activity is formed; stimulation of another point of the cortex (first pair) at this time causes excitation of these neurons. The increased excitability lasts for a long time after the end of tetanization and is revealed by stimulation of another cortical point a considerable distance away from it. This last stimulation, moreover, imposes its rhythm on the neurons of the focus of increased excitability.

Roitbak thus showed that during combined stimulation of two points of the cortex (tetanic stimulation of one point, comparatively infrequent stimulation of the other) rhythmic electrical activity arises, and the frequency of this rhythm is determined by the frequency of the infrequent stimulus, i.e., this stimulus imposes its rhythm on neurons of the focus of increased excitability. The additional stimulation can "control" the activity of the neurons in the focus of increased excitability and determine the rhythm of their excitation not only when it is combined with the tetanic stimuli producing these foci, but also in the period following the tetanic stimulation. The infrequent stimulation, when combined with the tetanic, does not excite that group of neurons until their excitability reaches a certain critical level at which it can be excited rhythmically by impulses reaching the cortex, regardless of their origin. The effect of stimulation depends not only on the preceding stimulation of that particular cortical point and the preceding stimulation of other cortical points, but also on the fact that the preceding stimulation of that cortical point had coincided with stimulation of some other cortical points.

In the oscillographic experiments described above, in which tetanic stimulation was applied to the cortical surface, a focus of increased excitability possessing, in Roitbak's words, the features generally regarded as characteristic of a dominant focus, was formed. "In fact, whenever we raise the excitability of a certain center," Ukhtomskii wrote, "and it is able to summate and to retain excitation, any fresh application of a random stimulus will evoke a response primarily in that center."

Essentially the same phenomenon was found by Beritov (1917) during local application of strychnine to the motor cortex, evoking rhythmic movements of the corresponding limb. These movements were increased by stimulation of many different areas of the cortex of both the same and the opposite hemisphere. Beritov explained this phenomenon, in which besides the center on which the stimulus acts, other centers whose excitability is increased under the influence of any intrinsic or extrinsic stimuli, also respond, by irradiation of excitation. The initial activity of the focus of increased excitability inhibits the response which usually arises to the particular stimulus used, although sometimes this response may appear (Beritov, 1910; Beritov, Bakuradze, and Narikashvili, 1937).

From an analysis of his own observations and of results obtained by Burns (1954) in experiments with an isolated strip of cortex, Roitbak concludes that if the somata of pyramidal neurons are in a state of catelectrotonus (resulting from anodal polarization of the cortex by a weak direct current), the excitability of the cells is increased and they undergo rhythmic excitation. The same phenomenon is observed in cortical neurons under the influence of relatively high-frequency excitatory impulses, and under natural conditions a focus of increased excitability is undoubtedly produced in this way.

The rhythmic character of excitation in the dominant focus was demonstrated in our laboratory by Grechushnikova (1963). Ukhtomskii (1925) himself, in an investigation with M. I. Vinogradov on the spinal frog, showed that a dominant focus can be formed by excitation of waning strength. Grechushnikova confirmed these findings in long-term experiments on rabbits. The rabbit's paw was first stimulated by an induction current at 12/min; the strength of the stimuli was then gradually reduced. The dominant focus was actually formed in the period following the waning stimulus. For 1 h or more, a motor response of the same limb appeared to indifferent nonrhythmic (continuous application of photic stimulation, tones, bells, buzzers) stimuli 20-40 sec in duration, and these movements were mainly in the rhythm of the waning stimulus applied at the beginning of the experiment. Hence it follows that the focus experiences excitation in a latent form and that this excitation is rhythmic in character.

When functional dominant foci such as these are formed it is difficult to say where the primary focus of excitation lies in the central nervous system. Whatever the case, the cortex participates in the formation of a dominant focus by a waning rhythmic stimulus, because its formation is accompanied by a change in the level of the cortical steady potential. These changes in cortical potential are reflected as periodic slow waves, with a closely similar rhythm to that of the waning stimulus which produced the focus. Periodic changes in the potential level took place most clearly after repeated formation of a dominant focus. As a rule, a change in the steady potential (an increase in its negativity) preceded the appearance of the rhythmic motor response (Grechushnikova, 1963).

4. Generators of Rhythmic Electrical Activity in the Human Cortex, Particularly the Motor Cortex, as Revealed by Correlation Analysis

The question of interaction between two rhythmically functioning foci in the central nervous system was partly considered in Chapter I. The correlation method reveals new details not apparent on visual analysis of interaction between the rhythms of activity of foci of excitation.

It is assumed that the rhythmic electrical activity of the human cortex is due to a series of generators. In fact, as correlation analysis of the human EEG has shown (Grindel', 1966), individual generators of rhythmic electrical activity exist both in the motor cortex and in the visual cortex, notably within the alpha-rhythm range. The interaction between these rhythm generators can now be examined.

There are several theories on generators of the alpha-rhythm. Some believe that there is a single generator (Garoutte and Aird, 1958; Dubikaitis and Dubikaitis, 1962), while others are of the opinion that several generators of alpha-rhythm exist, or they distinguish a rhythm of identical frequency, but with different properties, such as the rolandic or mu-rhythm (Walter, 1950; Esser and Bickford, 1950; Gastaut, 1952; Bekkering et al., 1957). By the use of cross-correlation analysis, the temporal relationships between the alpha-rhythms can be examined in different parts of the cortex and the degree of their correlation can be assessed quantitatively. The techniques of recording and analysis and the principle governing the action of the particular correlograph used are fully described by Grindel' et al. (1964). The alpha-rhythm of the occipital (O_1, O_2), central (C_1, C_2), and frontal leads (F_1, F_2)* was analyzed at rest, during continuous photic stimulation, and during muscular exertion — squeezing the fingers of both hands into a fist.

The following criteria were used to assess the correlograms: the mean frequency (f) and intensity of the periodic process, reflected in the coefficient of correlation between the power of the periodic (quasiperiodic) and random components ($C_{p/r}$), calculated for the segment of the correlogram from 0 to 1000 msec. The coefficient of cross-correlation (C_{cr}) and the time shift (TS) of the maximum of the cross-correlation function also were determined in cross-correlation analysis.

Autocorrelation analysis showed that the mean frequency of the period of the alpha-rhythm varied somewhat in these regions. The mean value of $C_{p/r}$ in all these groups showed no significant variation in the background. However, the periodicity of the process in the central region was higher than in the other two regions, especially in persons without a dominant alpha-rhythm in the EEG. The alpha-rhythm of the occipital, central, and frontal regions recorded on an analyzer, and the autocorrelation functions (ACF) of this alpha-rhythm and the background conditions, during continuous photic stimulation, and during squeezing the fingers into a fist are shown in Fig. 32. The highest mean frequency in the background and the greatest value of $C_{p/r}$ are found in the central region.

No significant changes in the mean frequency of the periodic process were found on the ACF during photic stimulation in any region of the cortex by comparison with the background. However, during photic stimulation significant differences were established between the mean frequency of the alpha-waves in the occipital and frontal regions and in the central and frontal regions (for the

---

*International scheme of EEG recording with "monopolar" leads, and using a common reference electrode on the ear.

4. GENERATORS OF RHYTHMIC ELECTRICAL ACTIVITY IN HUMAN CORTEX 163

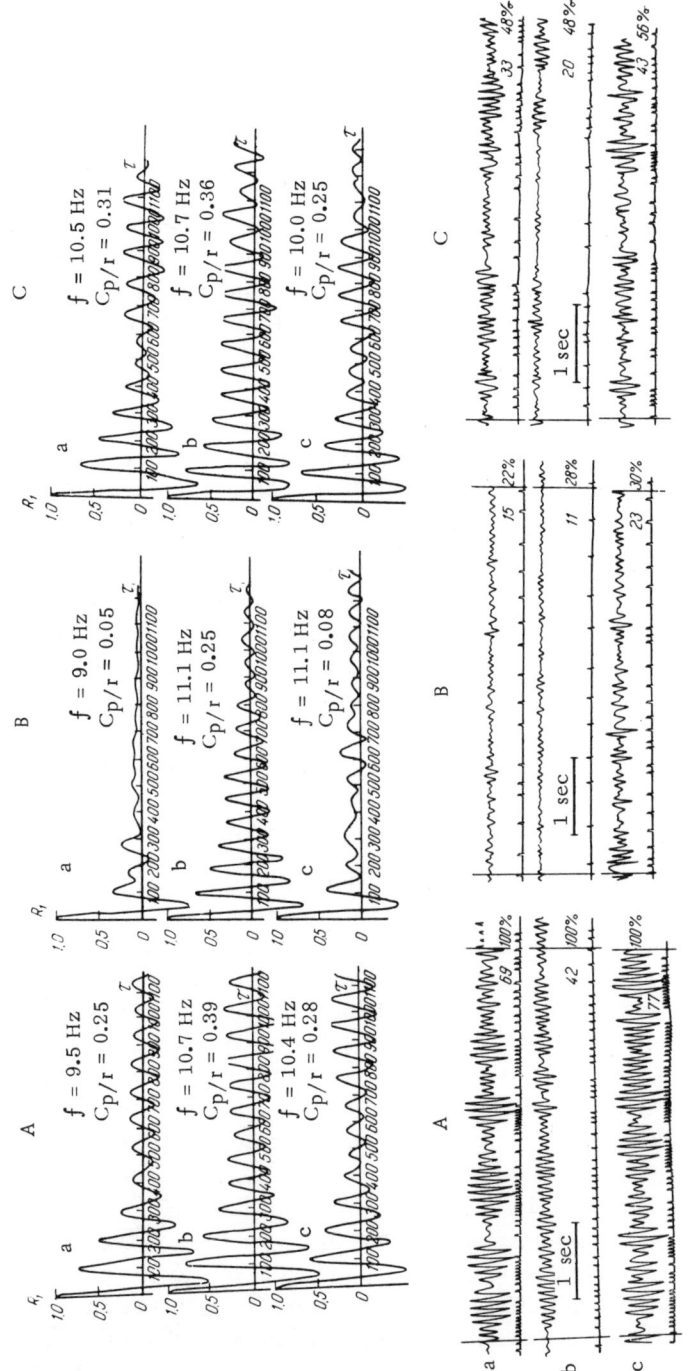

Fig. 32. Autocorrelation functions (above) and analyzer write-out (below) of alpha-rhythm of occipital (a), central (b), and frontal (c) regions of a healthy person (work of Grindel'). A) background; B) photic stimulation; C) squeezing fingers into a fist. Marker of integrator shown beneath each analyzer curve; readings of integrator given in absolute figures over 5 sec and in percentages; f) mean frequency of rhythm. C_p/r) coefficient of correlation between power of periodic and random components. τ in milliseconds.

occipital and frontal regions t = 2.3, P < 0.05; for the central and frontal regions t = 4.0, P < 0.01).

As Fig. 32b shows, photic stimulation yields various changes in ACF: the occipital lead shows a marked disturbance of the periodic process: $C_{p/r}$ fell from 0.25 to 0.05 with a slight decrease in the mean frequency. In the central region the change in $C_{p/r}$ was smaller, from 0.39 to 0.25 with a small increase in the mean frequency of the alpha-rhythm. In the frontal region the periodic process was considerably reduced: $C_{p/r}$ fell from 0.23 to 0.08. The ACF showed that during depression there is a marked difference in the rhythm of the process. During photic stimulation a disturbance of the rhythm (desynchronization) is observed in the occipital cortex, while in the central region a similar decrease in amplitude is unaccompanied by any disturbance of the periodic process, i.e., desynchronization does not take place (Grindel', 1965).

Investigation of the response of the alpha-rhythm to squeezing the fingers into a fist revealed a different pattern. In all 3 regions depression of the alpha-rhythm was observed during the period of squeezing, but the depression was less severe than during photic stimulation. Under the influence of muscular exertion no regular changes in the mean values of $C_{p/r}$ took place compared with the background, although in individual subjects definite changes in the periodicity and frequency of the alpha-rhythm took place, especially if a marked rolandic rhythm was present in the EEG.

Investigation of the cross-correlation function of the occipital alpha-rhythm compared with the central alpha-rhythm, and of the central and frontal alpha-rhythms showed a significant difference in the background between the mean values of the cross-correlation coefficients (C_{cr}) of these two pairs of regions C_{cr} for the occipital and central regions was 0.55; C_{cr} for the central and frontal regions was 0.71; t = 3.2, P = 0.01. There is also a noticeable difference in the shape of the cross-correlation functions (CCF). Predominance of the periodic component is more typical of the CCF of the occipital and central regions (mean value of $C_{p/r}$ 0.75), while the random component is typically predominant in the CCF for the central and frontal regions (mean value of $C_{p/r}$ 0.33). However, because of the great variability of $C_{p/r}$ for CCF of the occipital and central regions, the difference in the periodic process of these two CCFs was not significant. The mean frequency of the rhythm

4. GENERATORS OF RHYTHMIC ELECTRICAL ACTIVITY IN HUMAN CORTEX

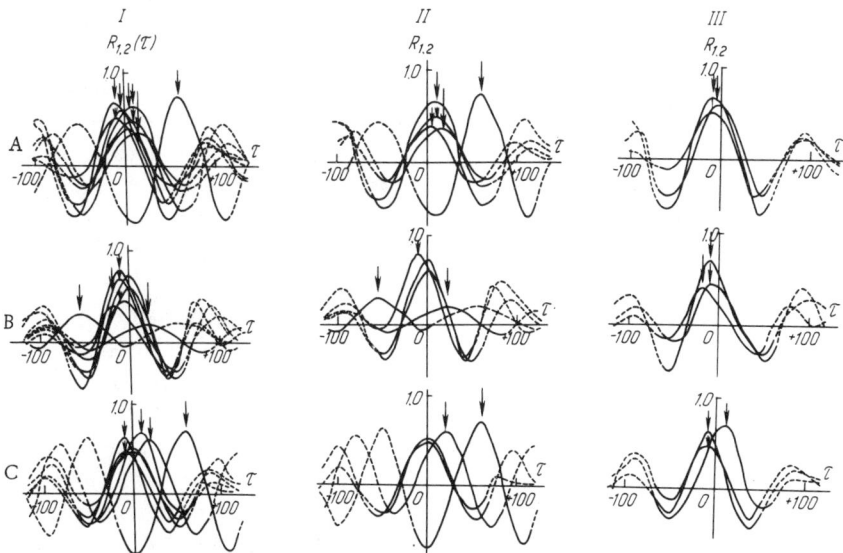

Fig. 33. Changes in shift of maxima of cross-correlation function of alpha-rhythm under the influence of stimulation (work of Grindel'). Superposed graph of CCF: I) of all subjects; II) of subjects with dominance of alpha-rhythm in EEG; III) subjects without alpha-rhythm in EEG. CCF of occipital and central regions; A) background; B) during photic stimulation; C) making fist. Arrows indicate maximum value of cross-correlation functions. τ) magnitude of shift of processes relative to each other (in milliseconds).

common to the occipital and central and to the central and frontal regions showed no significant difference.

The most regular findings were changes in temporal relationships on the CCF of alpha-activity during photic stimulation. Great variation in the temporal shift in the CCF of the occipital and central regions in the background can be seen in Fig. 33, I. Some changes are evidence of delay of the process in the occipital region compared with the central, while others, on the other hand, indicate delay in the central region. If the subjects are divided into groups depending on the character of the EEG, i.e., on the dominance or absence of a marked alpha-rhythm, the pattern of distribution of the TS is simplified: if a dominant alpha-rhythm present in the background, the TS indicates delay of the common process in the central region compared with the occipital (Fig. 33,

II); in the absence of a dominant alpha-rhythm the TS indicates delay of the process in the occipital region (Fig. 33, III). During photic stimulation, the TS in both groups is to the left, indicating delay of the process in the occipital region. This shows that the central generator of the alpha-rhythm during photic stimulation is dominant relative to the occipital generator. It can accordingly be considered that the rhythmic alpha-activity appearing in the occipital region, sometimes with increased, sometimes with reduced frequency, is not a true or, more precisely, is not only a true occipital alpha-rhythm arising from the occipital generator, but is to a greater degree a rhythm arising from the central generator. During clenching of the fingers into a fist, accompanied also by inhibition of the alpha-waves, the temporal relations between the occipital and central regions were similar to those in the background. Correlation between alpha-activity in the central and frontal regions, as Fig. 34 shows, is different from that between the occipital and central regions. The high values of C_{cr}, indicating very close correlation, will be noted; TS = 10-5 msec, and frequently its direction indicated delay of the process in the central region relative to the frontal.

The differences in behavior of the alpha-rhythm in different parts of the healthy human cortex revealed by correlation analysis indicate that several generators are present. Depending on the state of the brain and incoming afferent stimulation, their activity may change. Alternately they dominate or are dominated by the influence of other generators which are inactivated at that particular moment as the result of incoming afferent impulses. The changes in alpha-activity in the central region are unique. This zone of the brain is particularly interesting as an association zone, a zone of convergence of impulses of different modalities and manifestation of the nonspecific response (Gastaut, 1953; Kats, 1958; Rusinov, 1962; Puchinskaya, 1963). Responses of this region of the cortex to photic stimulation differ in the intensity of the periodic component from those of the occipital and frontal regions. During photic stimulation special relationships are established between the regions which differ from those in the background. The rhythm is less depressed than that in other regions, notably in the central region. With a decrease in intensity of the alpha-rhythm, or during depression, the periodic component in this region remains unchanged or may actually be intensified.

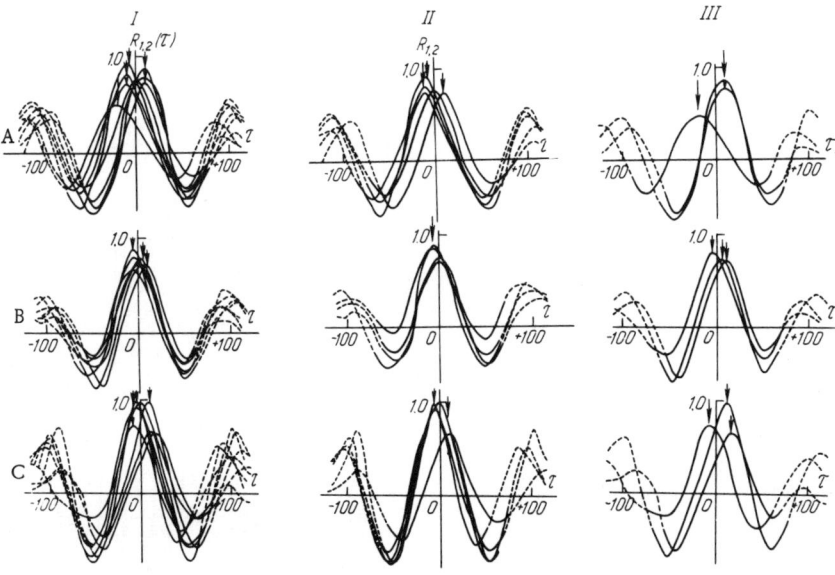

Fig. 34. Changes in shifts of maxima of cross-correlation functions of alpha-rhythm under the influence of stimulation (work of Grindel'). CCF of central and frontal regions. Legend as in Fig. 33.

This suggests that although photic impulses disturb the rhythm in the occipital and frontal regions, they do not do so in the central region, but maintain it. During the period of photic stimulation two mechanisms of action of the cortex evidently operate: the first mechanism depresses the amplitude of electrical activity and acts uniformly on all parts of the cortex investigated, while the second mechanism influences the synchronization and periodicity of the activity. Photic impulses spreading along the specific system into the visual cortex disturb the rhythm of the occipital region, and modify the correlation between the regions characteristic of the resting state, in which the occipital cortex plays a leading role. Impulses arising from photic stimulation reaching the central region along diffuse nonspecific pathways, on the other hand, strengthen its periodicity; a link is established between the two regions in which the central zone plays the dominant role. In persons whose EEG is not dominated by the alpha-rhythm (with an EEG reflecting a state of activation) the central region also plays a leading role

168   CHAPTER VI: SPECIFIC AND NONSPECIFIC THALAMO-CORTICAL SYSTEMS

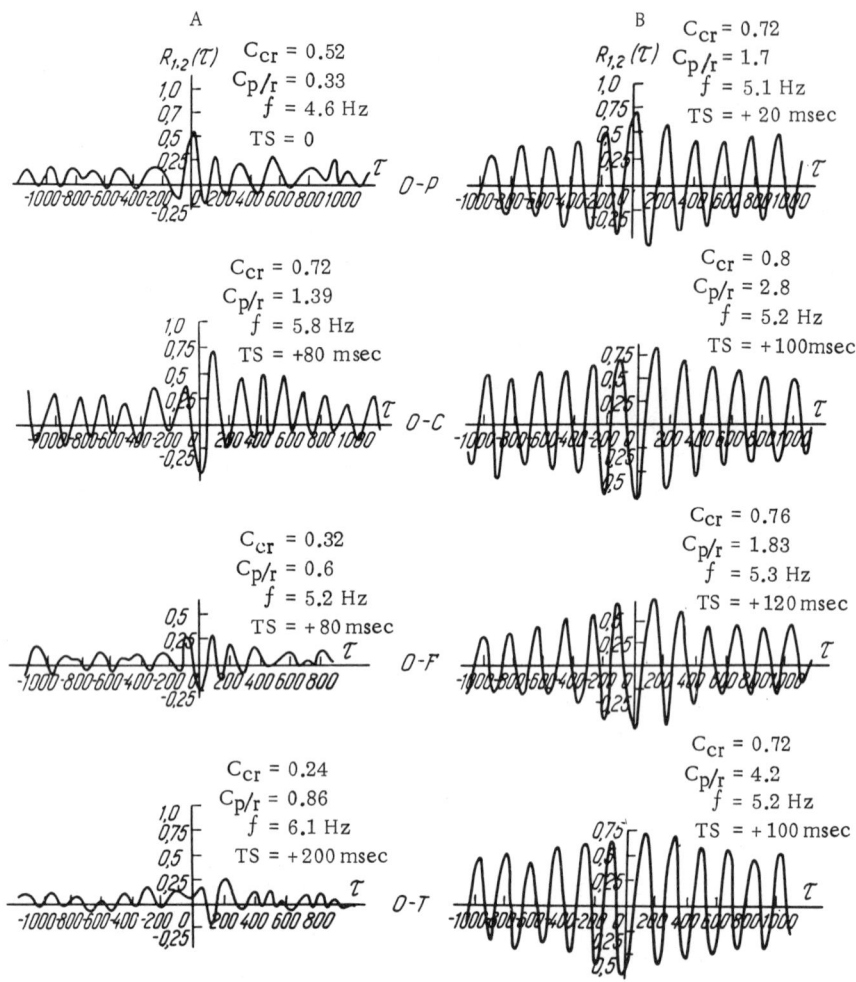

Fig. 35. Cross-correlation functions of theta-rhythm of occipital (O) region with theta-rhythm of the parietal (P), central (C), frontal (F), and temporal (T) regions of the left hemisphere of subject A, obtained from the EEG by means of a frequency analyzer; background (A) and during repetitive photic stimulation (B) at 6 Hz; $\Delta\tau$ = 20 msec (work of Boldyreva).

in the background state, in the absence of additional stimulation, as is shown by the temporal shift on the CCF (Grindel', 1966).

5. Correlation Analysis of Rhythmic Driving of the Human EEG by Flashes

By correlation analysis it is possible to determine the composition of the EEG reflecting a complex stationary random process and to distinguish from irregular waves activity which systematically maintains its periodicity.

It is possible to distinguish periodic components in a process by correlation analysis even if their amplitude is much less than the amplitude of the background waves of random origin. Since rhythmic driving by flashes is similar to a periodic process, the application of correlation analysis to its investigation is of great interest.

With repetitive photic stimulation the CCF changes significantly in parts of the brain investigated (Boldyreva, 1965). Besides background rhythms, the CRF of the theta-rhythm of test pairs of leads during repetitive photic stimulation (6 Hz) are shown in Fig. 35. In the first place, there is a marked increase in C_{cr} compared with the background in all CCFs examined, indicating an increase in the closeness of correlation between the various parts of the hemispheres during repetitive photic stimulation. A periodicity in rhythm with the flashes is detected in all CCFs shown. The ratio between the power of the periodic and random components ($C_{p/r}$), like the value of C_{cr}, is considerably increased in all pairs of leads examined.

If CCR for the analyzed pairs of the EEG of each individual subject are compared, the degree of synchronization of the activity of the occipital region with other regions can be estimated, i.e., a measure can be obtained of the functional interaction between different parts of the cortex during photic stimulation. The EEG of the occipital region showed the highest degree of correlation with the EEG of the parietal and temporal regions, and a lower degree with the EEG of the central and frontal regions. Repetitive photic stimulation had a marked effect on the temporal relationships between the EEGs of the regions tested.

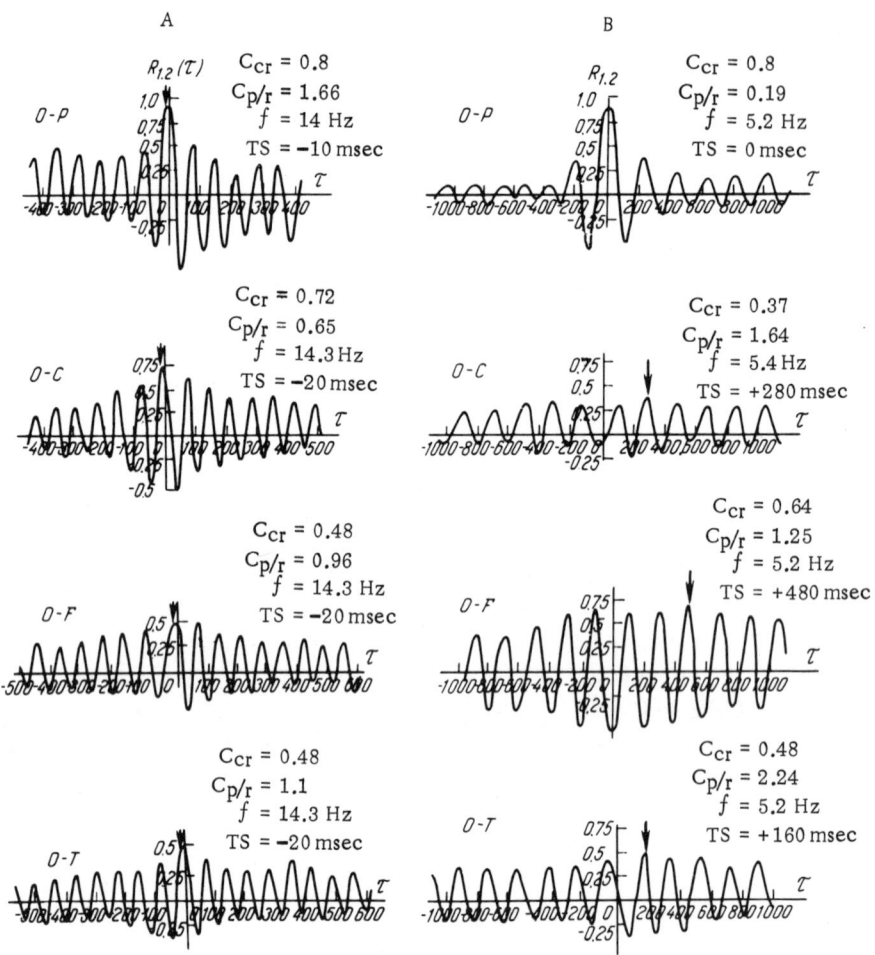

Fig. 36. Cross-correlation functions of test pairs of leads from the left hemisphere of subject M during photic stimulation at 15 Hz (A, analysis of separated beta-rhythm range, $\Delta\tau$ = 10 msec) and at 6 Hz (B, analysis of theta-rhythm range separated from EEG, $\Delta\tau$ = 20 msec) (work of Boldyreva). Legend as in Fig. 35.

5. CORRELATION ANALYSIS OF RHYTHMIC DRIVING OF THE HUMAN EEG

It is clear from the CCFs of the background activity in Fig. 35 that the temporal shift (TS) varies between the tested regions. In the CCF of the occipital and parietal regions the TS is absent, indicating that the EEGs of these regions are in phase. In the CCFs of the occipital and central and frontal and temporal regions the maximum of the cross-correlation function was shifted along the τ axis to the right. This indicates anticipation of the rhythm isolated from the occipital EEG by the periodic component of these parts of the brain. During rhythmic driving by flashes at 6 Hz, as the CCF in Fig. 35 shows, the magnitude of TS between the tested areas changes. The periodic activity, in rhythm with the flashes, in the occipital region anticipates that in the parietal region by 20 msec, in the central region by 100 msec, and in the frontal region by 120 msec. This successive increase in TS in this case is evidence that a periodic process following the rhythm of the flashes spreads from the occipital to the frontal region.

During photic stimulation the magnitude and direction of the TS varied considerably, and it was largely determined by the frequency of photic stimulation. The CCF of the theta-rhythm isolated from the EEGs of all tested regions during photic stimulation at 6 Hz and the CCF of the beta-rhythms during stimulation by flashes at 15 Hz are given in Fig. 36. In the background CCF of both theta- and beta-rhythms the periodic activity of the anterior regions anticipated that of the posterior regions; the magnitude of this anticipation was relatively small and varied from 10 to 60 msec. If the driven rhythm of flashes was low (6 Hz), the maximum of the cross-correlation functions of all CCFs of the theta-rhythm was shifted to the right along the τ axis. This suggests that the periodic activity bound to the rhythm of stimulation arises sooner in the occipital region than in the central and anterior parts of the brain. During stimulation at relatively high frequency (15 Hz) the maxima of the cross-correlation function of the beta-rhythm in the tested region were shifted to the left along the τ axis. In that case, periodic activity bound to the rhythm of flashes appeared sooner in the central and anterior parts of the hemispheres than in the occipital region.

There is a marked difference in the magnitude and character of the TS of the driven rhythm depending on the direction of spread of the process. If the rhythm spreads in the postero-anterior direction, there is a successive and considerable increase in TS from the occipital to the frontal poles. If the rhythm is delayed in the occipital region, the temporal shifts in the CCF are small (10-20 msec), and are closely similar for all pairs of regions tested. This behavior is evidently due to differences in the conduction paths of the afferent impulses evoked by photic stimulation to the regions studied. In the first case, the appearance of the rhythmic, driven response in the central and frontal regions is due, in all probability, to irradiation of the driving from the specific visual projection area in the occipital region. After their analysis of this phenomenon, Walter et al. (1949) put forward the hypothesis that the visual area is a generator system which transmits its signals to other areas of the cortex. If the bound rhythm in the central and frontal regions anticipates the rhythm of the posterior regions, the appearance of the rhythmic, driven response anteriorly is not secondary to the response of the visual area, but is evidently due to the arrival of impulses along the nonspecific afferent system. The possibility of flashes producing rhythmic driving in the anterior regions through activation of the nonspecific afferent system has been confirmed experimentally (Gastaut and Hunter, 1950; Hunter and Ingvar, 1953, 1955; Crighel, 1959; Sager, 1961).

The appearance of a rhythmic driving response in the EEG was accompanied by a marked increase in C_{cr} over the background level, indicating an increase in the degree of synchronization of electrical activity of the various cortical regions during afferent stimulation. These results are in agreement with those obtained by Livanov (1962a, b) when analyzing the spatial correlation of potentials by means of the electroencephaloscope.

During photic stimulation two types of spatio-temporal relationships were distinguished: a successive spread of the bound rhythm from the occipital region to more anterior parts of the hemisphere, manifested by an increase in TS in the occipito-frontal direction; the EEG rhythm of the fronto-central regions, bound to the rhythm of the flashes, anticipated the rhythm of the occipito-parietal regions.

In subjects in whom the modified rhythm spread from the occipital region to more anterior parts of the hemisphere, an alpha-rhythm which also spread in an occipito-frontal direction was recorded in the background EEG. In subjects in whose background EEG either the alpha rhythm was absent or it spread in a different manner over the cortex, the modified rhythm in the anterior and central parts of the hemispheres anticipated the rhythm in the posterior regions of the brain, driven in rhythm with the flashes.

This difference in temporal relationship of the bound rhythm in the various regions is evidently due to heterogeneity of the responsible mechanisms. Investigations along these lines in man confirm those of experiments on animals which show that excitation evoked by photic stimulation can be conducted to the central region of the hemisphere not only along cortico-cortical pathways arising in the visual area, but also along nonspecific pathways running directly to these parts via the reticular system.

The use of auto- and cross-correlation analysis to investigate the spatio-temporal behavior of the bound rhythm in the EEG appears very promising as a method of elucidating the mechanisms of integrative activity of the brain.

6. The Rhythmic Driving by Flashes when a Pathological Focus Is Present in the Brain

Boldyreva (1966), working in this laboratory, also investigated the response to rhythmic flashes for patients with organic brain lesions (brain tumors, head injuries). Whereas patients with a gross pathological focus of electrical activity (as delta-waves) show no photic driving in the affected hemisphere, patients with an irritative focus exhibit a more marked driving on the side of the lesion, and the driving, moreover, is usually manifested more strongly in the area of the pathological focus. The most interesting results in this connection are those obtained in patients with a focus in central or anterior regions and with no photic driving response in the occipital region. The clinical picture in these patients contained features of irritation of the brain stem. Flashes at 4-7 Hz (within the theta-rhythm range) evoked the clearest driving response in these patients in the anterior, and not in the posterior regions, where such driving to flashes usually occurs. Selective driving in anterior regions by flashes in the frequency range of the theta-rhythm evidently indicates that this response is linked predominantly with activation of the nonspecific system.

One of the distinguishing features of the central nonspecific region of the cortex, where nonspecific impulses converge, is the preservation, or even the strengthening, of its electrical rhythms, especially in response to repeated photic stimuli. During repetitive photic stimulation two types of spatio-temporal relationships can be distinguished by correlation analysis: a successive spread of the bound rhythm from occipital to anterior regions, reflected by an increase in TS of the maximum of the cross-correlation function in the occipito-frontal direction, which must evidently reflex the participation of a cortico-cortical mechanism in the formation of this response; second, anticipation of the photically driven rhythm by the fronto-central compared with the occipito-parietal regions. In the latter case, the appearance of the photically driven rhythm anteriorly is not secondary to the response of the visual cortex, but is evidently due to the arrival of impulses over the nonspecific afferent system.

The question of the relationship between the specific and nonspecific thalamo-cortical systems is intimately connected with the problem of convergence of impulses at the neuronal level.

Chapter VII

Convergence of Impulses on Neurons of the Motor Cortex, and the Motor Dominant

1. Functional Convergence and Polysensory Neurons of the Motor Cortex

Attempts have recently been made to classify the widespread convergence of impulses in the central nervous system. Although the validity of classifications based on other criteria is not denied, the investigation of dominant foci enables two types of convergence to be distinguished: a static convergence, structurally fixed by existing connections, and a dynamic convergence, functional and formed in the course of the response.

Functional convergence of impulses in the motor cortex is of considerable interest. Single unit activity in the motor cortex has been investigated by several authors, notably Phillips (1956, 1959, 1961), who was one of the first to describe the electrical activity of the large pyramidal neurons, which he identified by means of antidromic stimulation. Several features of the discharge of pyramidal neurons have been studied by means of extracellular and intracellular recording (Albe-Fessard, Buser, et al., 1953; Li, 1959; Phillips, 1961; Kuznetsov, 1963; Voronin and Skrebitskii, 1965; Voronin, 1966). A detailed investigation by Buser and Imbert (1964) showed that most neurons of the motor cortex are polysensory, i.e., that they respond to stimulation of all modalities (cutaneous, photic, and acoustic).

Nearly all the investigations of unit activity in the motor cortex cited above were carried out on immobilized or anesthetized animals. The question of the relationship between the motor responce and activity of the corresponding neurons thus remains to be investigated. In the few studies on nonimmobilized animals (Morrell, 1961; Vasilevskii, 1965; Voronin, 1966), no such comparison was made. Only Ricci, Doane, and Jasper (1957) have described changes in unit activity during CR movements. Their investigation revealed several types of changes in single unit activity in response to both the CS and US, accompanied by a defensive motor response.

To examine the phenomena taking place in experiments with the motor dominant, Sokolova and Lipenetskaya (1966), working in this laboratory, made a microelectrode analysis of neurons at that point of the motor cortex where dc polarization is usually applied. They analyzed 224 neurons, recorded extracellularly in unanesthetized, nonparalyzed rabbits. Of this total number, 109 neurons did not respond to electrodermal stimulation while the remaining 115 did. In the latter group, 78 neurons responded by an increase and 33 by a decrease in firing rate. Neurons responding by an increase in firing rate to electrodermal stimulation were divided, in turn, into two groups: monosensory neurons (22%), responding to electrodermal (adequate in this case) stimulation only, and polysensory neurons (78%), also responding to stimulation of other modalities (acoustic, photic).

Without dwelling on differences in the latent periods, frequency of the response, and other parameters of the monosensory and polysensory neurons in rabbits, differences in their relationship to the motor response must be mentioned. The monosensory neurons respond to movement by a grouped discharge, which corresponds to the beginning of the movement. Polysensory neurons also respond by an increase in firing rate to movement, but this increase begins before the movement and coincides with the beginning of the "arousal," activation, or desynchronization response, to use the various terms describing the response of depression in the EEG (Fig. 37).

The experiments of Sokolova and Lipenetskaya showed that the increase in discharge of the polysensory cells arises only when the

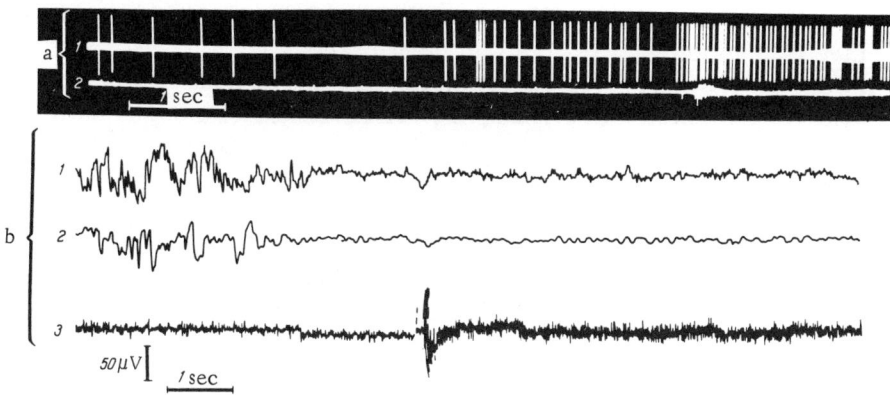

Fig. 37. Changes in activity of a polysensory (nonspecific) neuron during spontaneous movement (work of Sokolova and Lipenetskaya): a) simultaneous record of unit activity (1) and electromyogram (2); increase in frequency of spontaneous unit activity 2 sec before onset of action in electromyogram; b) concurrent ink-writer record; arousal reaction anticipates motor response by 2 sec; 1) EEG of motor cortex; 2) EEG of visual cortex; 3) electromyogram.

stimulus (nonspecific) evokes an arousal reaction in the EEG. The fact that polysensory neurons respond to stimulation of different modality by an increase in their firing rate only if the stimulus at the same time evokes an arousal reaction in the EEG was confirmed by Kalinin and Sokolova (1968) in their experiments with stimulation of the mesencephalic reticular formation of rabbits.

The presence of polysensory neurons in the motor cortex producing re-entrant excitation upon themselves or, more precisely, responding by activity to stimuli of different modalities, is an essential condition, although not sufficient by itself, for the response of the effector to nonspecific stimulation. The decisive condition for the motor response is that the neurons receiving convergent excitation attain a threshold level of activity, and send impulses to the effector whose area of representation in the cortex is undergoing dc polarization. If under these circumstances the nonspecific system acts, its activity will be exhibited locally: a movement will arise in response to the nonspecific stimulus, not in every limb but mainly, or even strictly, in one limb in response to stimulation of optimal strength.

Under the influence of a weak direct current acting on the surface of the motor cortex, as was mentioned above, a focus of excitation is formed, and under optimal conditions of polarization and afferent stimulation it will acquire the character of a dominant focus. The local nature of the response is manifested not only in the movement of one limb corresponding to the site of polarization, but also in the prolonged muscle tone in that same limb (Fig. 38). The prolonged tonic response reflects the formation of a focus of excitation in the central nervous system. The fact that a tonic response is obtained as an aftereffect of stimulation merely confirms the active role of the afferent stimulus itself in the formation of the focus. Voronin (1966), in his experiments in which single units of the motor cortex of the waking rabbit were polarized, showed that weak and, in some cases, subthreshold depolarization of the cell membrane is an adequate condition for responses to appear to nonspecific stimuli in a large proportion of neurons. By taking intracellular recordings from neurons of the motor cortex, Voronin found that the most characteristic features of spontaneous activity are continuous fluctuations in the level of the resting potential, indicating a sustained synaptic bombardment of the cortical neurons of the unanesthetized animal. In "partially intracellular" recordings, when the microelectrode evidently indents the cell membrane, penetrates partly into it, but does not penetrate inside the cell itself, changes in the membrane potential of the neuron can be recorded for long periods, sometimes up to 2-3 h.

By extracellular and intracellular recordings of responses to photic and acoustic stimuli in the motor cortex Voronin found that the motor area in rabbits, just as in cats, is a center of polysensory convergence. Many of the neurons respond not only to somatic stimulation, but also to stimuli of other modalities. With respect to changes in the responses with time, a group of neurons which either began to respond, or whose existing response was strengthened during repeated stimulation of the same or different types, deserves special attention. This group of neurons is interesting in connection with their possible role in the formation of temporary connections.

Burns (1954) postulated that the polarizing current evokes depolarization of cell bodies evidently in the region of the initial axonal segments of vertically oriented cortical neurons. Subthresh-

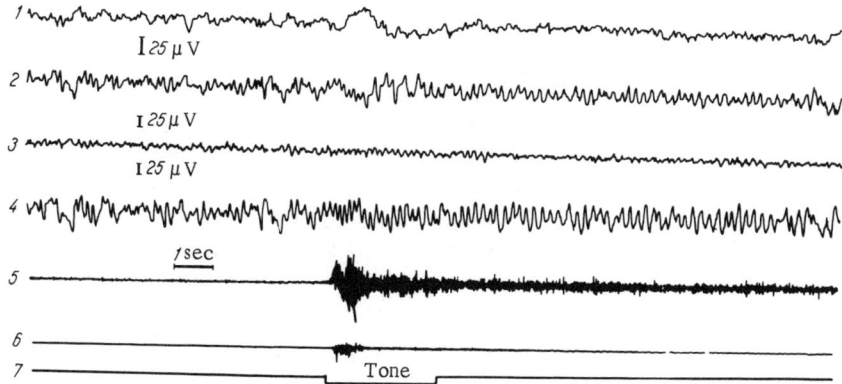

Fig. 38. Prolonged tone of muscles of the left forelimb as aftereffect of afferent stimulation in the presence of dominant focus in right forelimb area: 1) right motor cortex; 2) right visual cortex; 3) left mesencephalic reticular formation; 4) left thalamic reticular formation; 5) EMG of left forelimb; 6) EMG of right forelimb; 7) afferent stimulus (tone). Anodal polarization of right motor cortex, 5 μA (work of Kalinin).

old EPSPs, in the presence of artificial polarization of the membrane through an intracellular electrode, can cause motoneurons to discharge (Araki and Otani, 1955; Kostyuk and Semenyutin, 1961; Araki, 1965). The excitatory action of an outward current from an extracellular microelectrode (tip about 1μ in diameter) on background unit activity was demonstrated by Strumwasser and Rosenthal (1960). This effect was explained by assuming that a current entering a cell hyperpolarizes only a very small part of the membrane actually under the electrode, and neighboring parts of the neuron, including the initial segment, are depolarized by the outward current. In this last case the action of the current is similar to that of a depolarizing current through an intracellular microelectrode. In other words, additional depolarization of the membrane, however it is produced, can act as a stimulus to augment discharge of the neuron.

Having confirmed Burns's (1954) hypothesis that depolarization of the cell body is an adequate condition for responses to appear to nonspecific stimuli in a large proportion of neurons in the motor cortex of waking rabbits, Voronin showed, however, that the results of polarization of the cortical surface are different from those of polarization of single units. The percentage of cells responding to afferent stimulation during surface polarization is much greater than during polarization of single units, and in addition, in the latter case there are no responses to "nonspecific" stimuli after removal of the current polarizing the single neuron. Much evidence has been obtained to show that an anode on the cortical surface depolarizes not only the region of the cell bodies, but also in the zone generating fast prepotentials, which evidently lies close to the bifurcation of the dendrites. Intracellular recording has shown that one cause of the greater neural "elaboration" could be intensification of the background synaptic bombardment. Results obtained by intracellular recording also showed that changes in activity take place in neurons directly beneath the polarizing electrode mainly as a result of a change in the SPL, while in cells at a distance from the polarizing electrode a more important role is played by effects due to the modification of activity in neighboring units. The action of the current on presynaptic endings may also be a factor strengthening synaptic effects.

Voronin and Skrebitskii (1964) showed that many neurons of the motor cortex increase their spontaneous firing rate under the influence of anodal polarization of the cortical surface by a current of 3-10 μA. Neurons not responding to photic and acoustic stimuli begin to respond to these stimuli against the background of polarization. The effects continue for some tens of minutes after the current is removed.

By a combination of intracellular and "partially intracellular" recordings from neurons of the sensorimotor cortex of unanesthetized rabbits Voronin and Solntseva (1969) investigated trace changes in firing rate after removal of direct current passed through the recording microelectrode. Slow changes in firing rate, indicative of adaptation, were observed during the action of the current. After removal of the depolarizing current, afterinhibition occurred in all recorded neurons, and disconnection of the hyperpolarizing current was followed by afteractivation. The duration of these aftereffects increased with an increase in the duration and strength of the current and reached 10 sec for intracellular recordings. No significant changes in membrane potential level or changes in synaptic bombardment were found which could satisfactorily explain these aftereffects. In most cells, during the after-inhibition large depolarizing waves were observed; and despite the fact that they exceeded the threshold level observed before polarization, these waves did not evoke discharges. Voronin and Solntseva conclude from their investigation that this common type of effect following intracellular polarization is most probably due, not to interaction between the polarized neurons and neighboring units, but to the properties of the polarized neurons themselves. In the case of cortical neurons the adaptive properties of the membrane are mainly responsible for the aftereffects of intracellular polarization.

Voronin (1970) investigated the neurophysiological mechanisms of aftereffects of repetitive stimulation of the cortical surface on the cortical neurons and showed that in this case the increased cell response is based on postsynaptic potentiation (PSP) of excitatory synaptic connections, both to testing stimuli applied to the cortex through electrodes used for conditioning tetanization and (Voronin, 1971) to cortical and peripheral stimuli at some distance from the point of tetanization. The appearance of responses or strengthening of existing responses to hitherto ineffective or relatively ineffective cortical and peripheral stimuli form, in Voronin's opinion, a "dominant focus" as the result of preliminary tetanization; changes taking place under these circumstances at the single unit level can be regarded as the "cellular analog" of the dominant focus. In this case the main cause of appearance of unit responses is evidently an increase in the effectiveness of synaptic action.

Voronin, Skrebitskii, and Sharonova (1971) examined the data on convergence of stimuli of different sensorimotor modalities and concluded that three types of convergence can be distinguished depending on the type of single unit responses: those of type 1 are most characteristic of "associative" structures, those of type 2 of primary sensorimotor systems, and those of type 3 are most characteristic of "nonspecific" structures. However, neurons with different types of convergence can be found in virtually any structure. The types of convergence may change with the functional state and the preceding activity of the brain.

It has been pointed out that during conditioning, neurons which usually respond only to an adequate stimulus begin to respond to hitherto indifferent stimuli (Bureš and Burešova, 1965; Shul'gina,

1966; Vasil'evskii, 1966). This same phenomenon was also found by Kotlyar and Shul'govskii (1966) and by Shul'govskii (1967) when they investigated unit activity during the formation of polarization and CR dominance. Unit responses during formation of the defensive CR dominant are thus similar in this property to the dominant state produced in the cortex by anodal polarization.

Rabinovich (1963, 1967) considers that the factor determining the evolution of the principal neurons of the motor cortex from their original ability to respond only to adequate stimuli, through the phase of polysensory responses, to special monosensory responses to a CS, is motor reinforcement, i.e., the flow of proprioceptive impulses during the motor act. Accepting this view of the role of proprioceptive impulses in the genesis of polysensory neurons, Voronin showed that the effects of surface polarization of the motor cortex are reproduced by electrodermal stimulation of the animal's paw accompanied by a motor response. Neurons in the motor cortex indifferent to photic and acoustic stimulation begin to respond actively to these stimuli against the background of motor responses evoked by electrodermal stimulation, and also subsequently. Electrodermal stimulation of the animal's limb of threshold strength for motor responses evokes a marked increase in spontaneous activity and the appearance of responses, in the form of single spikes and discharges, to photic and acoustic stimuli.

The phenomena observed by Voronin (1966), Voronin and Skrebitskii (1964), Purpura and McMurtry (1965), and Landau et al. (1965), who found that anodal current causes depolarization of the cell bodies of deep neurons, in fact take place under conditions close to those of the dominant focus during anodal dc polarization. However, the situation created by direct current at the cortical surface is much more complex. Kuznetsova (1963) found that the sign and magnitude of the shift in cortical steady potential under the influence of anodal polarization by a weak direct current (1-10 μA) depend on the original SPL which, in the motor cortex of the waking rabbit (when measured relative to an electrode placed on the nasal bone), is in most cases electropositive. During anodal polarization under these conditions the level of its steady potential shifts towards negativity (Fig. 39). In these cases the assumptions

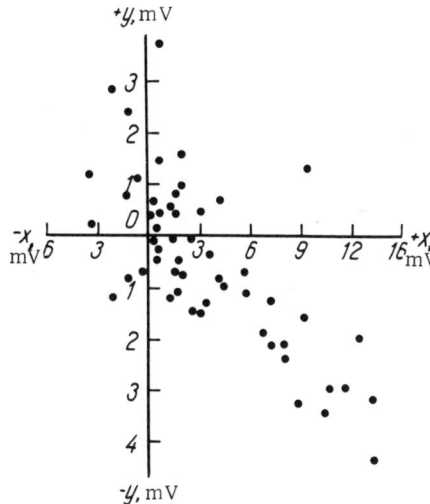

Fig. 39. Changes in potential of the rabbit motor cortex evoked by a single standard polarization as a function of the initial steady potential level (work of Kuznetsova). Abscissa, initial level of steady potential of motor cortex at beginning of polarization; ordinate, shift of potential after polarization for 15 min (results of 56 experiments on different rabbits).

of Phillips and Burns are unlikely to be valid. The anode does not directly excite the region of the pacemaker of the pyramidal cells, but it acts rather through synaptic mechanisms. Anodal polarization can restore the excitability of the bodies of nerve cells, peripheral nerves, and muscles after blocking by various methods. Research into this problem has been surveyed, although incompletely, by Narahashi (1964). A number of hypotheses have been put forward to explain this phenomenon. One, in particular, postulates that the restoration is due to repolarization. However, anodal polarization can abolish the block produced by cocaine, which does not depolarize (Lorente de Nó and Condoris, 1959), and also by certain alcohols, which hyperpolarize slightly. An interesting hypothesis was put forward by Narahashi, who suggests that the anode restores the normal membrane structure. Narahashi thereby supports the view of Tobias (1959) that the anodal current has a restorative effect by reintegrating the molecular structure of the membrane, which has been more or less disturbed by the action of blocking agents, notably those producing depolarization.

To discover the degree of "inertia" of the processes causing a shift of SPL of the motor cortex during its polarization, experiments were carried out on rabbits with repeated application of the current. An anode with current density 0.8 $\mu A/mm^2$ was applied at intervals of 5-10 min. The cortical potential thereupon steadily moved in the negative direction until it became weakly negative. A further application of the direct current at intervals gave two possible results. The first was transition from negative changes to positive, evidently reflecting the normal inhibitory response, while the second was a further increase in negativity with the appearance of very slow waves with a period of 6-7 min and amplitude up to 3 mV, accompanied by epileptoid activity. Paroxysmal movements began with the forelimb whose area of cortical representation was beneath the polarizing electrode, and spread gradually to all the muscles of the body (Kuznetsova, 1963). A normal dominant focus was thus converted into a pathological epileptogenic focus. The mechanism of choice between one of these possibilities is not yet clear, but a connection is definitely seen between the epileptogenic focus and the focus of excitation with dominant properties, just as a link is seen between the SPL in the focus of excitation and the character of the response — normal or pathological.

The very slow waves obtained in the cortex after repeated polarization by a weak direct current spread over the cortex just as when potassium chloride or weak pulses of a sufficiently strong current are applied to it (spreading depression).

A mathematical expression was obtained by Kuznetsova and her collaborators for the passage of a current through the polarized cortex, allowing for the resistivity and volume of the different cortical elements and intercellular spaces. Their calculations showed, as a first approximation, that the density of the current passing through the nerve cells is many times smaller than the net density of the current passing through the thickness of the cortex as a whole. Most of the current flows in the intercellular spaces, where it evidently produces certain changes in the synapses and in the chemical properties of the neuronal membranes.

2. Changes in Ultrastructure of the Motor Cortex following Polarization by Weak Direct Current

In a joint investigation of dominant foci, N. I. Artyukhina and I used the electron microscope to study changes produced by cortical polarization in the ultrastructure of an area of the motor cortex in rabbits before and after anodal polarization with a weak direct current in an experiment lasting 6 days, and after the formation of a dominant focus. Despite the considerable similarity between the ultrastructure of most synapses of the control and experimental rabbits, in the latter group of animals some synapses differed from the control. In such cases the number of synaptic vesicles was reduced, and in certain synapses they were concentrated in small numbers and only close to the presynaptic membrane. At the same time, particles of a dense substance accumulated in the synaptic cleft, where they formed dark bands lying perpendicularly to the membrane. Considerable diffusion of the dense particles under the postsynaptic membrane also was observed. These ultrastructural features of the synapses reflected their increased functional activity. Furthermore, in some synapses, besides a decrease in the number of synaptic vesicles, the synaptic membranes lost their discrete linear outlines and became loose in structure and confluent with the synaptic cleft. These membranes, it can be assumed, were more permeable to the particles of the diffusing substances. The synaptic vesicles were often deeply recessed into the presynaptic membranes with this type of structure.

Data in the literature on changes in synaptic structure by activity are extremely contradictory. De Robertis (1958) investigated changes in synaptic endings under the influence of electrical stimulation and found, in his own words, "dramatic" changes. During prolonged stimulation at 400 Hz, these changes consisted, in particular, of a marked decrease in the number of synaptic vesicles. In response to less frequent stimulation (100 Hz) there was an increase in the number of synaptic vesicles in the terminal axons, a process of "mobilization" of chemical mediators (Curtis and Eccles, 1960; Smirnov, 1968).

Changes in the ultrastructure of synapses observed by Artyukhina under the influence of anodal polarization are in agreement with the results obtained by de Robertis (1958). They indicate that during anodal polarization of the cortical surface in waking rabbits, as well as the changes described above, synaptic activity is intensified in some axo-dendritic synapses in which there is increased consumption of the contents of the synaptic vesicles.

* * *

Most neurons of the motor cortex of the waking rabbit which respond by increased activity to afferent stimulation are polysensory. Anodal polarization leads to the formation of a focus of excitation which can be used as a model to investigate the mechanisms of dynamic functional convergence of impulses from stimuli of different modalities. If the direct current is optimal in strength, impulses from afferent stimuli to which the orienting reflex was previously extinguished exceed the threshold level, travel to the effector organs, and evoke a motor response of the corresponding limb. Among the factors determining convergence in the net effect of anodal polarization of the surface of the motor cortex in the central nervous system, when hitherto indifferent stimuli evoke a motor response, there is also increased activity of some of the synapses formed by terminal axons and dendrites.

3. General Scheme of Processes in the Central Nervous System during Anodal Polarization of the Motor Cortex

In its general features, the pattern of events taking place in the central nervous system during polarization of the motor cortex is as follows. A change in cortical SPL takes place only for a very short distance over the surface and into the depth of the cortex, thus creating a focus of local excitation with increased excitability. Judging from the spread of the potential during polarization of the cortical surface by weak anodal current (density ~ 1 $\mu A/mm^2$), the focus is localized in the first two cortical layers, and it mainly involves apical dendrites of pyramidal neurons, interneurons and, possibly, glia. The anodal action on subjacent excitable structures varies not only with the strength of the current, but also with the functional state of the structure on which it acts. Afferent impulses of different modalities, spreading diffusely through the reticular system are not transmitted to pyramidal neurons in the motor cortex, despite the fact that the reticular system is connected either directly or through interneurons with their apical dendrites. Usually this connection of the reticular system with the pyramidal neurons is inhibited. Just as the anode restores conductance of a nerve in a focus of alteration (the model of an "inhibitory" synapse), it disinhibits the connection between the reticular system and the pyramidal neurons. Judging from the presence of two optimum strengths of the direct current, disinhibition takes place in the following order. Since the connection of the reticular system directly with the apical dendrites of the pyramidal neurons, it must be assumed, is simpler than that through interneurons, this connection is disinhibited first, and by a weaker current. This response corresponds to the first optimum, when movement of the limb whose area of cortical representation is being polarized appears for the first time in response to hitherto indifferent stimuli. With an increase in current this direct connection of the reticular system with the apical dendrites is inhibited, corresponding to the first minimum of the curve. With a further slight increase in the current the second connection of the reticular system with the apical dendrites (via interneurons) is disinhibited, possibly through the participation of the glia. This corresponds to the second optimum.

The appearance of an additional potential on the descending phase of the dendritic potential during cortical stimulation against the background of the weak direct current used in the experiments with the dominant focus is evidence in support of the view that interneurons participate in the conversion of the focus of excitation into a dominant focus, when the same response is obtained regardless of the place or type of stimulation. Results show that this connection of the reticular system with the pyramidal neurons is inhibited, in turn, after a further increase in current. Such is the possible course of events in the central nervous system during polarization with current between 1 and 10 μA and in response to afferent stimuli of different modalities. It is interesting to note that with a further increase in current the characteristic response of the dominant focus can be obtained once again, but it does not usually appear during the stimulation but as an aftereffect or after removal of the polarizing current, indicating an inhibitory response from the action of the strong direct current. It is possible that with a further increase in current other connections with the pyramidal neurons are disinhibited. As Ryabinina has shown, a response of dominant type is obtained also by cathodal polarization in the layer of large pyramidal cells, and under these conditions the anode is inhibitory.

The dominant focus formed by polarization by a weak, direct current can be used as a model to investigate the excitable and conducting structures of the central nervous system participating successively in the response of a focus of excitation. The same motor response of a dominant focus during polarization by a current of between 1 and 10 μA is produced by two different structures in turn, and this evidently guarantees its reliable performance.

The dominant focus, like any system with feedback, as was mentioned above, may be one of the mechanisms of memory, as is clear from experiments with cortical polarization with a pulsating direct current.

Two types of feedback, positive and negative, are known in the physiology of the central nervous system. The distinguishing feature of these types of feedback is that in the course of responses one type can change into the other. The dominant focus is formed as the result of summation of incoming impulses with the local stationary excitation. This is one of the mechanisms of formation of the dominant focus. Impulses can travel along recurrent collaterals of the same axons and, consequently, in this case a positive type of feedback is created. Later, the same focus of excitation, having attained its culmination, begins to respond to the same impulses, traveling along recurrent collaterals, not by the dominant reaction, but by its inhibition, thus exhibiting the second, negative type of feedback. The feedback intervenes in the course of the reflex associated with the dominant focus and changes its character depending on the SPL in the focus of excitation itself. Systems with feedback are of two types: in the first the feedback takes place without delay, while in the second delay is present and the effect of the feedback is exhibited only after a certain time has elapsed.

Ukhtomskii considered the dominant focus to be stable as the result of "inertia," which he regarded as one of the principal properties of the dominant focus. In fact, the dominant focus is stable because of feedback, and the physiological mechanism of the dominant is a system of regulation with feedback.

By their physiological properties, by their degree of complexity, simple foci of excitation differ from foci of excitation which become dominant. If a dominant focus is present there is a system of several feedbacks, controlling the maintenance of the dominant role of the reflex by which the focus of excitation responds to incoming impulses regardless of the place or type of stimulation.

The dominant focus with a system of feedbacks maintaining stability at a certain relative level is a reflex regulator in which the adjustment of the feedback is varied in the course of the response: a slave system is formed.

Feedbacks in the form of axon collaterals to their own cells are one type of feedback "to self" in the central nervous system. Feedback with delay is more widespread, and in the central nervous system it is represented either by interneurons along the path of the collateral to its own neuron, or as groups of neurons closed "on themselves" to form a circle of neurons. Any synapse along the path of a collateral to its own neuron represents a simple type of delay.

An interesting phenomenon was found by Eccles (1965) in his investigation of reciprocal inhibition. Simple reciprocal inhibition has been most adequately studied in cases of discharge of a motoneuron. The more intensive the discharge of the motoneuron the stronger the negative effect through axon collaterals and Renshaw cells. As a result of a more detailed investigation Eccles reached the following general conclusion: the action of a feedback system is a nonspecific limitation of the activity of motoneurons irrespective of their function. Negative feedback from impulses to any single muscle does not act selectively on the population of motoneurons inervating that muscle. In any population of motoneurons a focus of excitation in the center is least affected by the action of feedback, but this focus will be contracted or concentrated by the pressure of discharges of more weakly excited motoneurons at its periphery. In other words, the negative feedback system inhibits weakly discharging motoneurons. Eccles evidently feels that this conclusion applies to motoneurons in a focus of excitation which have not yet reached their culmination.

What is the possible role of delay in a system of neurons? Does differentiation of stimuli take place where memory is stored? These questions, like others, are connected with the role of delay in a system of neurons and they still await an experimental answer.

Results relating not only to the formation of a dominant focus, but also to its conversion to a new rhythm, when there is an apparent divergence between activity of the upper layers of the cortex and activity of the pyramidal cells, raise the question of how far the activity of deeper structures is reflected in surface recordings. The results relating to reorganization of the dominant focus suggest that structures in which trace processes are stored for long periods exist in the cortex.

Not only the state of polarization of structures in which a given process took place, but also the state of polarization of the structures connected with them functionally is important for the manifestation of trace activity. Trace phenomena in the cortex could be intensified or even restored by polarization of the medial thalamus by a current of a certain strength. During polarization of the medial thalamus (2 μA) repetitive movements in a certain rhythm reappear. The configuration of the contractions becomes more inert: they do not die away for a long time after application of acoustic or photic stimuli.

As mentioned above, application of a current with a steady component provides the best condition for creating a dominant focus. Trace processes, on the other hand, depend fully on the state of the dominant focus for their manifestation: the more marked the dominant focus, the longer and the more regular the trace processes. Polarization of nerve cells or, more precisely, a change in the level of polarization up to a certain optimum, facilitates the storage of traces.

Since in polarization experiments the steady potential level changes first, the following conclusion seems likely: a change in steady potential toward a certain optimum facilitates the manifestation of memory. The region of the apical dendrites is the principal site of generation of the cortical steady potential. Recent observations have emphasized the role of apical dendrites in the formation of temporary connections. The interesting prediction was made by Bekhterev (1896) that the first layer of the cortex in general presents the most favorable conditions for connecting activity.

Conclusion

The dominant focus is a temporarily dominant reflex system controlling the activity of neural centers at a given moment; the influence of particular cortical foci on particular efferent systems under these circumstances is variable in time, and the "center" loses its role as an apparatus with a special functional significance. The ability to form dominant foci of excitation is a general property of the central nervous system. Diffuse waves from stimuli of different modalities excite all centers which at that moment are sufficiently excitable, but a dominant focus is formed only in those centers which can summate excitation. The transmission of traces from one moment to another must play a vital role in this process (Ukhtomskii, 1926).

Foci of excitation in the central nervous system are essential to all activity of the organism. When they become dominant they have a decisive influence on the course and outcome of a response which is taking place. The dominant focus is Ukhtomskii's "constellation," formed as a system during current activity of the organism at all levels of the central nervous system, in its different parts, but with a primary focus in one part and with a variable role for functions of individual components of the system. In other words, individual components of a given constellation may acquire a different functional role in connection with other constellations when participating in other activities. The dominant focus is based on the mechanism of a summation reflex, but it is in fact a more complex system: its physiological mechanisms are of a higher order than those of a simple summation reflex. The latter does not possess such high "inertia" as the dominant focus. Reciprocal relationships are not established with a dominant focus through reflex arcs fixed anatomically once and for all. At the beginning of its formation, the CR is similar in its physiological mechanisms to the dominant focus, but later it becomes basically different. The simplest dominant focus is formed at all levels of the central nervous system, and the presence of the cerebral cortex is not essential to its formation. Before a focus becomes dominant it passes through the stage of a summation reflex. The CR is initially a dominant focus before it becomes a CR. Such is the relationship between the physiological mechanisms of the summation reflex, the dominant focus, and the CR.

Investigation of two rhythmically operating foci in the central nervous system (swallowing and respiration) showed that when the swallowing center begins to acquire the properties of a dominant focus and reciprocally to inhibit the other rhythmically operating center, a focus of stationary excitation is formed in the dominant center itself. The gradual increase in tone of the swallowing muscles at this time is the peripheral reflection of stationary excitation of the swallowing center. The inhibited dominant focus can be disinhibited by stimuli of different modalities. The ability of the dominant focus to become disinhibited is one of its characteristic properties.

The principal mechanisms of formation of the connection in the initial period of formation of a conditioned swallowing reflex are induction relationships (external inhibition) and summation in the center of the US; the center receiving impulses from the US changes its functional state during formation of the temporary connection and it becomes a focus of summation. The time factor is one of the parameters determining the course of a response which is in progress. The effect of the CS is opposite in character depending on whether the stimulus is applied in the first or second half of the interval between swallows. In other words, the functional state of the various structures of the "working constellation," as it changes under the influence of stimulation, other conditions being the same, determines the outcome of the response to the stimulus.

The dominant focus is only one stage in the functional evolution of the focus of excitation in the central nervous system: not every focus of excitation becomes a dominant. The dominant is one form of activity of foci of excitation in which the focus is established at a certain level of stationary excitation enabling the summation of excitation reaching the nervous system, and at the optimum rhythm of operation for the particular conditions so that it is predominantly this focus which responds to the external stimulus and all other active foci are correspondingly inhibited. During rhythmic operation of the swallowing center, Wenckebach's phenomenon is found and is identical in its pattern with the same phenomenon in the human heart and the phenomenon obtained experimentally in the animal heart. It is due to the changing relationship between the spreading excitation and the focus of stationary excitation.

The presence of slow, long potentials in the cortex has now been demonstrated by several investigators. The amplitude and direction of potentials longer than 1 sec in duration depend on the different techniques used in the investigation, but the actual fact that slow potentials exist in the cortex is not disputed. In response to afferent input, fields of steady potential, differing both in origin and in direction, arise in the cortex both in the projection zone and elsewhere. If the stimuli are relatively weak, positive fields predominate and many changes take place in specific regions, while with an increase in the stimulus, negative changes become preponderant and many changes in the cortical SPL take place in the anterior regions. Without acceptance of the stationary long-lasting excitation as a normal factor in the activity of the central nervous system it would be difficult to understand the physiological mechanisms of formation of the dominant focus and of the temporary connection.

Slow potentials are not products of fusion and coincidence of individual waves of excitation, but the reflection of focal, long-lasting activity. The stationary local potential sometimes rises, sometimes falls under the influence of stimulation — it calibrates. Slow changes of steady potential recorded in some cases in the cortex reflect stationary excitation. Like fast waves, they reflect the process of excitation arising in a different form, namely as a slow stationary process. Just as the fast changes of potential reflect an ongoing wave of excitation, slow changes of potential reflect longer lasting changes of excitation. This "monistic" view of the relationship between excitation and its electrical component assumes the existence of foci of long-lasting excitation, reflected electrographically by slow potentials, as normal and extremely important factors in neural activity. This view also assumes that not only the action potential, but also the slow potential reflect an active physiological state.

Considerable changes in the cortical SPL take place during formation of the temporary connection. For defensive conditioning in rabbits with photic stimulation, changes of several hundred microvolts and from several seconds to tens of seconds in duration are recorded in the SPL from the cortical surface. During conditioning the shift of SPL undergoes changes not only in distribution over the cortex, but also in direction. Before pairing CS and US, the photic stimulation (CS), after extinction of the orienting reflex to it, evokes no changes in SPL. The first changes in response to photic stimulation as a consequence of pairing appear in the visual area and are negative in sign. However, changes in SPL, now positive in sign, subsequently become diffuse. With an increase in the number of pairings, although the changes in SPL become diffuse they now become negative again, and they lose their diffuse character only gradually, becoming localized in the motor and visual areas.

The concentration of slow potentials in the cortex (shifts of SPL) with stabilization of the temporary connection is a characteristic feature of the slow potentials during conditioning. Another characteristic feature is their "conditional nature." Slow potentials recorded in the phase of generalization to all stimuli appear in the cortex, with stabilization of the temporary connection and the formation of differentiation, only in response to the CS and not in response to the stimulus to which differentiation is formed. Local negative changes in SPL during formation of the temporary connection reflect an active state of the structures recorded, for the CR takes place against their background, and with extinction of the CR and the formation of differentiation, these local changes in SPL gradually decrease and disappear completely. The changes in SPL in this case simply disappear and are not replaced by changes with an opposite sign, reflecting an inhibitory state, as might be expected.

Two more facts deserve attention. Changes in SPL, unlike other phenomena in the dynamics of cortical electrical processes, do not disappear during formation and stabilization of the temporary connection in response to a large number of combinations of

CONCLUSION

conditional and unconditional stimuli. On stabilization of the CR in response to the US the SPL is more marked in the projection zone of the CS than of the US. For example, in the defensive CR of a rabbit to photic stimulation, the SPL to electrodermal stimulation is more marked in the occipital cortex than in the sensorimotor cortex. If the degree of activity is judged from the intensity of the change in cortical SPL, it still remains unknown which of these foci play the more active role in the stabilized temporary connections.

Without denying the possible role of postsynaptic potentials in the genesis of slow cortical potentials, their nature can be explained by a change in the membrane properties of the cortical neurons themselves. Extracellular electric fields generated by the neurons may play a role in interneuronal functional connections.

The results concerning changes in SPL, their changes during formation of a temporary connection, their stability during its stabilization, and their disappearance with extinction and the formation of differentiation are all evidence to support the hypothesis of the leading role of stationary fields, i.e., of slowly changing potentials in the closure of the temporary connection. That the local character of the SPL changes when the temporary connection is stabilized confirms the focal character of cortical activity in the constellation of structures participating in the temporary connection. Experiments in which artificial foci of excitation are formed by cortical polarization with a weak direct current, evoking a local change of SPL and leading to the formation of a dominant focus, so that the corresponding reflex is obtained in response to hitherto indifferent stimuli, give experimental proof of the validity of the hypothesis that processes reflected electrographically as shifts of steady potential play a leading role in the formation of temporary connections.

Anodal polarization (1-5 μA, current density of the order of 0.8 $\mu A/mm^2$) of the motor cortex of the nonimmobilized and unanesthetized rabbit at the projection of one of the limbs creates conditions under which afferent stimuli (photic, acoustic, tactile) to which the orienting reflex has previously been extinguished evoke a motor response of the corresponding, or mainly of the corresponding limb. This response is easily inhibited. All that is necessary to inhibit the dominant focus is to increase the strength of the direct current. As the current is gradually increased, the resulting response passes through two maxima of activation and two minima of depression between the range from 0 to 10 μA. The existence of these two optimal strengths of current for obtaining a motor response to hitherto indifferent stimuli suggests that in order to obtain a motor response to indifferent stimuli there must be two structures at the site of polarization which play a direct part in the formation of the dominant focus and facilitate the passage of impulses from these stimuli to the focus of polarization, and from thence to the effector. The distribution of potentials during weak polarization of the rabbit motor cortex shows that polarization affects mainly the surface layers (I and II). The coincidence of the distribution of potential over the cortical surface for a current of optimal strength (2 μA), reaching its lowest value at a distance of 4 mm from the point of polarization, and the extent of spread of the apical dendrites, which is also 4 mm, is interesting.

Bearing in mind the distribution of potential and the characteristic changes in shape of the dendritic potential when the strength of the polarizing current is increased, it can be postulated that the apical dendrites play the dominant role in the organization of the first optimum, while interneurons or glia are concerned with the organization of the second optimum. Polarization of the nervous system by weak direct current is not only a method of producing a model of the dominant focus, but it can also be a method of investigating the order of involvement of neural structures in the response to stimulation.

Recently the attention of physiologists has increasingly been turned to the possible role of the glia in the fundamental functions of the central nervous system. The following circumstances direct attention to the possible role of the glia in the closure of temporary connections.

The brain consists of two tissues: neurons and glia. The glia are mobile, but the laws of their movement are not yet known. Electron-microscopic studies picture the brain as being packed with glia. The empty spaces between the neurons seen in sections under the light microscope do not exist. All free space between the neurons is filled with glia. There is evidently a definite pattern of distribution of the glia. The question arises, what determines this pattern? The relations between the sheath of Schwann and nerve tissue in the peripheral nerves is precisely known: the axon grows only into the sheath of Schwann which directs its growth.

The sheath of Schwann is the peripheral glia. Possibly in the central nervous system the glia have a similar structure which also directs growth of the axon and its collaterals. Neurons can give off processes in all directions, but the vector of their growth is determined in the central nervous system, just as in the peripheral nervous system, by the glia. This view has already been expressed by Pribram (1962). This may well be one of the factors contributing to the general process of formation of temporary connections and a mechanism of memory.

It is also essential to discover the reason why the glia are arranged so as to enable a temporary functional and, perhaps, structurally fixed and constant connection between neurons. An electrostatic field is evidently established between foci of stationary excitation, bringing the interacting neurons into the same functional state. This field is manifested as slow electrotonic effects, long-lasting potentials, and shifts in SPL. The glia may serve as conductors of slow effects, reflected electrographically by slow changes of potential. Axon collaterals of an afferent neuron grow out and lengthen between the glial cells, and either an ephaptic or a direct contact connection is established. Since the intercellular space between the neurons, as electron-microscopy shows, does not exceed 200 A, only a very small change in their length is implied.

Whether the glia are in fact arranged along certain lines in the field formed between foci of excitation has not yet been shown experimentally. There is some evidence of the orderly arrangement of the glia when a pathological focus is present in the central nervous system. In particular, this has been shown in the case of a focus caused by implanted electrodes (Aleksandrovskaya, 1962). Investigation of the morphological changes in the tissue structures of the central nervous system after implantation of electrodes (tungsten electrodes, 150-170μ in diameter, insulated with varnish except at the tip) has shown that they are phasic, depending on the time after implantation of the electrode. The whole process is completed by the development of gliosis near the tip of the electrode and a generalized diffuse reaction of the microglia. Extensive mobilization of the astroglia and, in particular, of the microglia takes place. The reaction of the astrocytes is more local. The microglia, on the other hand, react diffusely and in parts of the brain further from the electrode. The fact that glial cells accumulate in large numbers near the tip of an electrode implanted in the brain recalls the observations of Tasaki and Chang (1958) who showed that astrocytes in tissue culture respond to stimulation by generating slow electrical activity, and also the depolarization effect discovered by Kuffler et al. (1966) in the glial cells during the passage of impulses in the amphibian optic nerve.

These investigations have brought new problems in the physiology of the central nervous system, especially in the field of the mechanism of formation of new connections.

A dominant focus in the motor cortex is stabilized not only by impulses of different modalities traveling along the pathways of the reticular system directly, or via interneurons, to the pyramidal cells, but the specific cortical area of the system receiving the reinforcing stimuli also plays a part.

The foci themselves can be subdivided, depending on the degree of their complexity, into simple foci of excitation, foci acquiring the character of a dominant and activated by any stimuli, and foci forming complex systems of temporary connections responding only to adequate stimuli.

Like any system with feedback, the dominant focus is a mechanism of memory, as is clear from experiments in which the cortex is polarized by a pulsating direct current. Preservation of the basic properties of the focus of excitation — its rhythmic nature, a focus of manifestation, stability at a certain development, and disinhibition — are characteristic of the trace phenomena associated with a dominant focus arising from polarization. During inhibition the cortical focus preserves its imposed rhythm in a latent state, and it is manifested after the inhibition. The view is widely held in the literature that only short-term memory can be explained by the circulation of nerve impulses, for a long circulation must inevitably cause fatigue, but this view is no longer tenable. It fails to take into account the inhibitory response which, during prolonged circulation, can develop sooner than fatigue. Consequently, there is no basis for rejecting prolonged activity of circulating impulses in a ring of neurons as a possible mechanism of long-term memory.

A polarized system of nerve cells can follow the parameters of the stimulus precisely and can thus store a model of the stimulus in its memory. Deeper neuronal structures in the cortical dominant focus retain the old rhythm longer than structures under the direct influence of cortical surface stimulation, and from which the electrocorticogram is recorded.

During inhibition, not only is the previously bound rhythm preserved, but "learning" may evidently continue, for the intensity of the dominant phenomena after removal of the direct current producing the inhibition depends on the number of test stimuli applied against the background of inhibition. There is also evidence to show the importance of the time factor to the quality of the assimilation of a rhythm by the central nervous system, for motor responses of the limb are qualitatively better a comparatively long time after than immediately after removal of the stimulus.

The physiological mechanisms of conversion of a focus of excitation into a dominant focus, when the same response is obtained regardless of the place and modality of stimulation, are connected with the diffuse spread of excitation but, at the same time, with its local manifestations and also with the dynamic convergence of impulses. Convergence of impulses may be of two types: one determined by existing anatomical connections, the other functional and dynamic, formed in the course of the reflex response on the appearance of the dominant focus.

Polarization of the medial thalamus by a weak direct current after the creation of a cortical dominant focus considerably increases the reflex movements evoked by indifferent stimuli. Additional polarization of the mesencephalic reticular formation has the same effect on the cortical dominant focus. Polarization of the specific thalamic nuclei reinforces the cortical dominant focus only if the thalamic nucleus which is polarized is concerned with the transmission of impulses from the particular stimuli used: additional polarization of the lateral geniculate body strengthens the dominant only in response to photic stimulation. To strengthen a motor dominant, i.e., to make it more stable, to evoke a larger number of limb responses and to strengthen them in response to photic stimulation, the additional polarization by a weak direct current must be applied to the medial geniculate body. Depending on the strength of the current applied, optimal excitation or depression both of the specific and of the reticular system are produced, but there is an essential difference in the final effect on the cortex. Polarization of the specific thalamic nuclei by direct current does not evoke diffuse changes in the cortex.

The focus of excitation produced in the hypothalamus by a weak anodal current (3-10 μA) possesses the main properties of a dominant: acoustic and photic stimuli to which the orienting reflex has first been extinguished evoke changes in respiration and cardiac activity. By forming a dominant focus in the hypothalamus, the animal's blood pressure can be changed. The blood pressure rises slowly during polarization of the hypothalamus. After cessation of the current polarizing the hypothalamus, the phenomena associated with the dominant focus persist for several hours or, in some cases, for several days.

At the moment of reinforcement of the dominant focus by photic or acoustic stimulation, changes take place in the hypothalamic electrical activity. These changes differ depending on the state of the focus. In the first stage of formation of the focus afferent stimuli slow the rhythm so that it approximates the rhythm of respiration. As the dominant focus becomes stabilized the amplitude of the waves increases and slower waves become predominant. The stronger the dominant focus during hypothalamic polarization, the lower the amplitude of the cortical waves at the time of reinforcement of the focus by afferent stimuli. In the absence of a dominant focus, i.e., when there are no changes in the autonomic responses to stimulation, there is no decrease in amplitude of the cortical potentials. Electrical activity of the cortex and hypothalamus is not restored until 40-50 min after removal of the direct current, bringing the dominant focus into a state of inhibition.

Polarization of individual layers of the rabbit's motor cortex by a current of at least 0.5 μA revealed the different roles of each layer in the formation of the motor dominant. For polarization of the lower cortical layers the clearest results are observed if the tip of the polarizing electrode is in layer V. The dominant then arises only with cathodal polarization. This cathodal effect is evidently due to the current leaving the pacemaker region of the pyramidal cells through the excitable membranes.

There is no consistent response in the amplitude of the evoked potentials to photic stimulation during the recruiting response. Stimulation of the medial thalamus evoking recruiting responses, judging by its effect on evoked potentials recorded from the cortical surface, may have either a facilitatory or an inhibitory action on the specific afferent system. This effect depends on the functional state, on the background, on the level of excitation of the structure stimulated. In response to simultaneous electrical stimulation of the medial thalamus and application of flashes, a two-way action is obtained: not only the evoked responses to the photic stimulus but the recruiting response itself is changed. The mutual influence of these two thalamic systems when they are stimulated simultaneously can be taken as evidence of the electrotonic character of this phenomenon.

Impulses evoking a nonspecific response in the cortex act diffusely, but the response arises in that region of the cortex whose level of excitation is raised. The fate of the final effect of the afferent stimulus evoking the nonspecific response is decided in the cortex by the mechanism of action of the focus which has assumed dominant properties. The nonspecific response to photic, acoustic, tactile, and electrodermal stimulation is moved toward the focus, i.e., there is a selective manifestation of the nonspecific potential. The fact that the nonspecific response moves during repeated stimulation in the direction of the primary projection area of the specific system does not confirm the view that the interaction between the two thalamic systems is competitive, but it points rather to their coordinated function and to a dynamic relationship between them: when a focus of excitation is present one system reinforces the other.

Comparison of the theory of the dominant focus with the theory of automatic control invites consideration of Fel'dbaum's (1960, 1966) theory of dual control. The dual control is responsible for the study of an object and also for its control. In complex automatic systems, the theory of control has to contend with some degree of initial indeterminacy, and it must try to solve the problem of optimal control of objects despite the incompleteness or even absence of initial information. This possibility is based on the use of adaptation and self-instruction in automatic systems, which reduce the initial indeterminacy through the use of information obtained actually during the controlling process. By adaptation is meant a change in the magnitude of the parameters, in the structure of the system, and also, possibly, in the controlling actions on the basis of information obtained during control, in order to attain an optimal state of the system despite the initial indeterminacy and the changing working conditions (Tsypkin, 1965). Optimalization is a central problem both in automatic control and in the formation of a dominant focus when a simple focus of excitation is converted into one with dominant properties in the central nervous system.

To produce a dominant focus by polarization, the current applied must be of optimal strength, which differs for different animals and varies in the course of the same experiment. In other words, there must be a change in the steady potential level in the group of neurons forming the focus, a shift of optimal degree for summation of impulses from stimuli of different modalities. The afferent stimuli must also be optimal in their parameters of strength and frequency. Afferent impulses reach the cortical focus along two systems — specific and nonspecific — along the fibers of the reticular system. The elucidation of the relationship between these two systems is essential for the theory of communication in the central nervous system. As the dominant focus becomes established the diffuse system loses its nonspecific features but is manifested increasingly locally, concentrating influence in the focus, and thus becomes highly specific as regards its functional role. The physiological mechanisms of this concentration of effect of the nonspecific system in the dominant focus is reciprocal inhibition, which acts as a system of negative feedback primarily on less strongly stimulated neurons, which will also includes neurons not in the focus of excitation.

The dominant focus is a reflex system including in its functional organization not only the primary focus, but also the specific cortical projection areas and corresponding subcortical structures of the systems receiving the afferent stimuli reinforcing the focus.

The results of correlation analysis of the electrical activity of the brain reveal definite spatio-temporal relationships between the processes responsible for the concerted activity of individual parts of the brain during its normal function. In particular, they demonstrate the existence of several simultaneously acting cyclic processes, differing in their rhythm and spatio-temporal characteristics, thereby indicating that different structures are involved in these cycles. The same cortical structure (the point which is analyzed) may be connected simultaneously with other cortical points at different frequencies. In the literature on connections in the sensory systems it is stated (Fessard, 1964) that segments of pathways transmitting information to different cells often are used jointly. This joint use of the same line limits the potential capacity of the system as regards the transmission of simultaneous items of information ("engaged lines"). The results of cross-correlation analysis of the human EEG point to simultaneous connection between a given cortical point and other points at different frequencies and thus evidently raise the question of "engaged lines" in relation to intracortical connections.

Afferent stimuli modify the correlation functions of electrical activity, create new spatio-temporal relationships between different parts of the cortex, and reveal the degree of participation of specific and nonspecific brain systems in the mechanism of responses.

The use of mathematical analysis and, in particular, of autocorrelation, cross-correlation, and spectrum analysis in the study of electrical activity and the changes in its various forms during the creation of dominant foci is a promising method of obtaining further information on the mechanisms of formation of new connections at the higher levels of the central nervous system with the ultimate purpose of controlling them.

Bibliography

Adrian, E. D., 1931, J. Physiol. (London), 72:132.
Adrian, E. D., 1932, J. Physiol. (London), 75:26.
Adrian, E. D., 1942, J. Physiol. (London), 100:159.
Adrian, E. D., and Buytendikj, F. J. J., 1931, J. Physiol. (London), 72:121.
Adrian, E. D., and Matthews, B. H., 1934, J. Physiol. (London), 81:440.
Airapet'yants, É. Sh., and Balakshina, V. L., 1933. Trudy Leningrad. Obshch. Estestvoisp., 62:141.
Aladzhalova, N. A., 1960, "Infraslow rhythmic waves of potential in the cortex and subcortical structures and factors influencing them," in: Problems in Electrophysiology and Electroencephalograph [in Russian], Moscow—Leningrad, pp. 213-219.
Aladzhalova, N. A., 1962, Slow Electrical Processes in the Brain [in Russian], Izd. AN SSSR, Moscow.
Albe-Fessard, D., Buser, P., and Fessard, A., 1953, "Analysis of the electrical activity of the sensory motor cortex of the cat," XIX International Physiological Congress. Abstracts of Communications, Montreal, p. 348.
Aleksandrovskaya, M. M., 1962, "Morphological changes in the brain of animals after implantation of electrodes," Dokl. Akad. Nauk SSSR, 143(6):1442
Allison, T., 1962, Electroenceph. Clin. Neurophysiol., 14(3):331.
Anokhin, P. K., 1956, "Role of the reticular formation of the brain stem in conduction of unconditioned excitation in the cerebral cortex," Proceedings of the XX International Congress of Physiologists [in Russian], Izd. AN SSSR, Moscow.
Anokin, P. K., 1957, Fiziol. Zh. SSSR, 43(11):1072.
Anokin, P. K., 1962, "The specific action of the reticular formation on the cerebral cortex," in: Electroencephalographic Investigation of Higher Nervous Activity [in Russian], Izd. AN SSSR, Moscow, pp. 241-251 (Suppl. 13, Electroenceph. Clin. Neurophysiol., 1960).
Araki, T., 1965, "The effect of strychnine on the postsynaptic inhibitory action," 23rd International Congress of Physiological Sciences, Vol. 4, Tokyo, p. 96.
Araki, T., and Otani, T., 1955, J. Neurophysiol., 18:472.
Arduini, A., Mancia, M., and Mechelse, K., 1957, Arch. Ital. Biol., 95(2):127.
Arshavskii, I. A., 1956, Uspekhi Sovr. Biol., 41(2):193.

Asratyan, E. A., 1937, "Anatomical and histological basis of the conditioned-reflex activity of higher animals," in: Physiology of the Central Nervous System [in Russian], Izd. Akad. Med. Nauk SSSR, Moscow, 1953, pp. 147-161.
Asratyan, E. A., 1938, "The switching principle in conditioned-reflex activity," in: Physiology of the Central Nervous System [in Russian], Izd. Akad. Med. Nauk SSSR, Moscow, 1953, pp. 162-167.
Asratyan, E. A., 1958, Zh. Vyssh. Nervn. Deyat., 8:305.
Asratyan, E. A., 1960, in: Central and Peripheral Mechanisms of the Motor Activity of Animals [in Russian], Izd. AN SSSR, Moscow.
Asratyan, E. A., 1963, "The conditioned reflex and related phenomena," in: Philosophical Problems in the Physiology of Higher Nervous Activity and Psychology [in Russian], Izd. AN SSSR, Moscow, pp. 323-357.
Bancaud, J., Block, V., and Paillard, J., 1953, Rev. Neurol., 89(5):399.
Barlow, J., 1964, "Computer techniques in EEG analysis," Electroenceph. Clin. Neurophysiol., Suppl. 20, p. 31.
Bekhterev, V. M., 1906, Fundamentals of the Science of Brain Functions [in Russian], No. 6, p. 853.
Bekkering, D. H., Kuyper, J., and Storm van Leeuwen, W., 1957, Acta Physiol. Pharmacol. Neerl., 6:632.
Bennett, M. V. L., Aljure, E., Nakajima, Y., and Pappas, G. D., 1963, Science, 141:262.
Bennett, M., Aljure, E., Pappas, G., and Nakajima, Y., 1967, J. Neurophysiol., 30(2):180.
Bennett, M., Gimenez, M., Pappas, G., and Nakajima, Y., 1967, J. Neurophysiol., 30(2):236.
Bennett, M., Nakajima, Y., and Pappas, G., 1967a, J. Neurophysiol., 30(2):161.
Bennett, M., Nakajima, Y., and Pappas, G., 1967b, J. Neurophysiol., 30(2):209.
Beritashvili, I. S., 1937, Fiziol. Zh. SSSR, 22:755.
Beritov, I. S., 1910, "Reciprocal innervation of skeletal musculature on local strychnine poisoning of the spinal cord," Trudy Obshch. Estestvoisp. SPb, 41(2):245.
Beritov, I. S., 1917, Russk. Fiziol. Zh., 1:12.
Beritov, I. S., 1932, Individually Acquired Activity of the Central Nervous System [in Russian], Tbilisi.
Beritov, I. S., Bakuradze, A. N., and Narikashvili, S. P., 1937, "Phenomena of excitation and inhibition in the central nervous system," Trudy Inst. Fiziol. im. I. S. Beritashvili, 3:173.
Beritov, I. S., and Nivinskaya, O., 1925, Med.-Biol. Zh., 4.
Beritov, I. S., and Roitbak, A. I., 1955, Zh. Vyssh. Nervn. Deyat., 5(2):173.
Beritov, I. S., and Roitbak, A. I., 1957, "The nature of central inhibition," in: Gagra Colloquia [in Russian], Vol. 2, Izd. Akad. Nauk Gruz. SSR, Tbilisi, p. 175.
Bishop, G. W., and Clare, M. H., 1953, J. Neurophysiol., 16(1):1.
Bodian, D., and Bergman, R. A., 1962, Bull. Johns Hopkins Hosp., 110(2):78.
Boldyreva, G. N., 1965, "The use of correlation analysis to assess topographical features of the rhythm-binding response to flashes in the human EEG," in: Mathematical Analysis of Electrical Phenomena of the Brain [in Russian], Nauka, Moscow, pp. 29-41.

Boldyreva, G. N., 1966, Zh. Vyssh. Nervn. Deyat., 16(4):684.
Bonnet, V., 1958, J. Physiol. (Paris), 50(2):163.
Bonvallet, M., Dell. P., and Hiebel, G., 1954, Electroenceph. Clin. Neurophysiol., 6:119.
Borisova, E. I., and Rusinov, V. S., 1940a, Klin. Med., 18(7-8):68.
Borisova, E. I., and Rusinov, V. S., 1940b, Klin. Med., 18(7-8):82.
Borisova, E. I., and Rusinov, V. S., 1949, Fiziol. Zh. SSSR, 35(2):216.
Brazier, M., and Casby, J., 1952, Electroenceph. Clin. Neurophysiol., 4:201.
Bremer, F., and Terzuolo, C., 1953, Arch. Internat. Physiol., 61:86.
Bremer, F., and Terzuolo, C., 1954, Arch. Internat. Physiol., 62:157.
Brodal, A., 1960, Reticular Formation of the Brain Stem [Russian translation], Medgiz, Moscow (Thomas, Springfield, Ill., 1957).
Brookhart, J., Arduini, A., Mancia, M., and Moruzzi, G., 1958, J. Neurophysiol., 21:499.
Bureš, J., 1957, Electroenceph. Clin. Neurophysiol., 9:121.
Bureš, J., and Burešova, O., 1965, J. Neurophysiol., 28(4):641.
Burns, B. D., 1954, J. Physiol. (London), 125(3):427.
Buser, P., and Imbert, M., 1964, "Sensory projections in the cat motor cortex," in: Theory of Communication in Sensory Systems [Russian translation], Mir, Moscow, pp. 214-231 (Rosenblish, W. A. (Editor), Wiley, New York, 1961).
Caspers, H., 1961, Electroenceph. Clin. Neurophysiol., 13(4):651.
Caspers, H., 1962, "Die Veränderungen der corticalen Gleichspannung und ihre Beziehungen zur senso-motorischen Aktivität (Verhalten) bei Weckreizungen am freibeweglichen Tier," XXII International Congress of Physiological Sciences, Vol. 1, Part 1, Leiden, pp. 443-447.
Caspers, H., 1964, "Relations of steady potential shift in the cortex to the wakefulness-sleep spectrum," in: Brain Function, Vol. 1, UCLA Forum Med. Sci., Los Angeles, pp. 177-214.
Caspers, H., and Schulze, H., 1959, Pflüg. Arch. ges. Physiol., 270(2):102.
Chang, H., 1951, J. Neurophysiol., 14:1.
Chang, H. T., 1962, "Some data on changes in excitability of cortical and subcortical neurons and the possible role of these changes in conditioned reflex formation," in: Electroencephalographic Investigation of Higher Nervous Activity [in Russian], Izd. AN SSSR, Moscow, pp. 69-78 (Suppl. 13, Electroenceph. Clin. Neurophysiol., 1960).
Clare, M. H., and Bishop, G. H., 1955, Electroenceph. Clin. Neurophysiol., 7(1):486.
Cowan, M., and McDonald, R., 1965, Nature, 207:530.
Crighel, E., 1959, Cercetari asupra reactivitatie cortical, Bucharest.
Curtis, D., and Eccles, J. C., 1960, J. Physiol. (London), 150(2):374.
Danilova, N. N., 1964, Zh. Vyssh. Nervn. Deyat., 14(1):9.
Davis, P., 1939, J. Neurophysiol., 2(6):494.
Dejerine, J., 1901, Anatomie des centres nerveux, Vol. 11, Paris, p. 1.
Dempsey, E. W., and Morison, R. S., 1942, Am. J. Physiol., 135:293.
Desmedt, J. E., and La Grutta, G., 1957, J. Physiol. (London), 136:20.
Doty, R., 1951, Am. J. Physiol., 166:142.
Dubikaitis, Yu. V., and Dubikaitis, V. V., 1962, Biofizika, 7(3):345.

Durinyan, R. A., 1964, Basic Features of the Central Organization of Visceral Afferent Systems. Doctoral Dissertation, Moscow.
Dusser de Barenne, J. C., and McCulloch, W., 1938, J. Neurophysiol., 1:69.
Dzugaeva, S. B., 1958, Zh. Vyssh. Nervn. Deyat., 8(5):942.
Eccles, J. C., 1959, The Physiology of Nerve Cells, Moscow, IL (Johns Hopkins Press, Baltimore, 1957).
Eccles, J. C., 1965, "The control of neuronal activity by postsynaptic inhibitory action," 23rd International Congress of Physiological Sciences, Vol. 4, Tokyo, pp. 84-95.
Eccles, J., Kostyuk, P., and Schmidt, R., 1962, J. Physiol. (London), 161(2):258.
Elinson, A., 1896, On the Vasomotor Nerves of the Retina, Kazan'.
Enomoto, T. E., 1959, Electroenceph. Clin. Neurophysiol., 11:219.
Esser, R. A., and Bickford, R. G., 1950, Electroenceph. Clin. Neurophysiol., 2(2):231.
Euler, C., 1950, J. Cell Comp. Physiol., 36:333.
Eyzaguirre, C., and Kuffler, S. W., 1955, J. Gen. Physiol., 39(1):121.
Ezrokhi, V. L., 1967, Electrophysiological Investigation of the Neuron and System of Two Neurons Treated with Stychnine or Novocain (Crustacean Stretch Receptors). Candidate's Dissertation, Moscow.
Ezrokhi, V. L., 1968, Biofizika, 13:86.
Ezrokhi, V. L., 1969, Neirofiziologiya, 1(3):309.
Ezrokhi, V. L., 1970, Neirofiziologiya, 2(3):321.
Feindel, W., and Gloor, P., 1954, Electroenceph. Clin. Neurophysiol., 6:389.
Fel'dbaum, A. A., 1960, Avtomatika i Telemekhanika, 21(9):1240.
Fel'dbaum, A. A., 1966, Fundamentals of the Theory of Optimal Automatic Systems [in Russian], Moscow, Nauka.
Fessard, A., 1964, "Role of neuron nets of the brain in the transmission of sensory information," in: Theory of Communication in Sensory Systems [Russian translation], Moscow, Mir (Rosenblith, W. A. (Editor), Wiley, New York, 1961).
Filimonov, I. N., 1959, "The reticular formation in the process of evolution of the central nervous system," Proceedings of the 9th Congress of the All-Union Society of Physiologists, Biochemists, and Pharmacologists [in Russian], Vol. 3, Minsk, pp. 43-48.
Florey, E., and Florey, E., 1955, J. Gen. Physiol., 39:69.
Fritsch, G., and Hitzig, E., 1870, Arch. Anat. Physiol. Wiss. Med., 37:300.
Furukawa, T., and Furshpan, E. J., 1963, J. Neurophysiol., 26(1):140.
Galambos, R., 1961, Proc. Nat. Acad. Sci. (Washington), 47(1):129.
Garoutte, B., and Aird, R., 1958, Electroenceph. Clin. Neurophysiol., 10(2):259.
Gastaut, H., 1952, Rev. Neurol., 87(2):176.
Gastaut, H., and Hunter, J., 1950, Electroenceph. Clin. Neurophysiol., 2(3):263.
Gastaut, Y., 1953, Rev. Neurol., 89:382.
Gerard, R. W., 1953, Scientific American, 189(3):118.
Gerasimov, V. D., Kostyuk, P. G., and Maiskii, V. A., 1964, Biofizika, 10(3):447.
Gerasimov, V. D., Kostyuk, P. G., and Maiskii, V. A., 1965, Fiziol. Zh. SSSR, 51(12):1434.
Glees, P., 1944, J. Anat. (London), 78:47.

Goldring, S., 1963, "Negative steady potential shift which lead to seizure discharge," in: Brain Functions, University of California Press, pp. 215-236.
Goldring, S., and O'Leary, Y., 1951, J. Neurophysiol., 14:275.
Goldring, S., and O'Leary, Y., 1954, Electroenceph. Clin. Neurophysiol., 6(2):201.
Goldring, S., and O'Leary, Y., 1957, Electroenceph. Clin. Neurophysiol., 9(4):577.
Goldring, S., O'Leary, Y., and Huang, S., 1958, Electroenceph. Clin. Neurophysiol., 10(4):663.
Goldring, S., O'Leary, Y., Winter, D., and Pearlman, A., 1959, Proc. Soc. Exp. Biol. (New York), 100:429.
Goldring, S., and O'Leary, Y., 1960, Fed. Proc., 19:612.
Golikov, N. V., 1933, "Functional changes in nerve on local strychnine poisoning," Trudy Leningrad Obshch. Estestvoisp., 62:33.
Golikov, N. V., 1950, Physiological Lability and Its Changes in Fundamental Nervous Processes [in Russian], Leningrad.
Granit, R., and Henatsch, H. D., 1956, J. Neurophysiol., 19:356.
Grechushnikova, L. S., 1962, "EEG changes in the presence of a motor dominant created by repetitive decreasing stimulation," Trudy Inst. Vyssh. Nervn. Deyat. Akad. Nauk SSSR, Seriya Fiziol., 7:33.
Grechushnikova, L. S., 1963, "Long waves of potential during creation of a dominant focus by repetitive decreasing stimulation," in: Nervous Mechanisms of Conditioned-Reflex Activity [in Russian], Izd. Akad. Nauk SSSR, Moscow, pp. 190-199.
Grechushnikova, L. S., 1964, The Defensive Dominant and Its Relationship with Conditioned Reflexes, Candidate's Dissertation, Moscow.
Grindel', O. M., 1965, "The role of correlation analysis in evaluation of the human EEG," in: Mathematical Analysis of Electrical Phenomena in the Brain [in Russian], Nauka, Moscow, pp. 15-28.
Grindel', O. M., 1966, Fiziol. Zh. SSSR, 52(10):1176.
Grindel', O. M., 1966, Frequency and Correlation Analysis of the Human Electroencephalogram Under Normal Conditions and in Focal Brain Lesions, Doctoral Dissertation, Moscow.
Grindel', O. M., Boldyreva, G. N., Burashnikov, E. N., and Andreevskii, V. M., 1964, Zh. Vyssh. Nervn. Deyat., 14(5):745.
Grindel', O. M., Boldyreva, G. N., and Andreevskii, V. M., 1965, "Auto- and cross-correlation analysis of the human EEG under normal conditions and in focal brain lesions," in: Bionics [in Russian], Nauka, Moscow, pp. 415-425.
Grindel', O. M., and Filippycheva, N. A., 1959, Zh. Vyssh. Nervn. Deyat., 91:545.
Grindel', O. M., Kandel', E. I., and Raeva, S. N., 1962, Vopr. Neirokhir., 6:23.
Gumnit, R., 1961, Electroenceph. Clin. Neurophysiol., 13(6):889.
Gustson, P. P., 1964, Cortico-fugal Effects in Two Systems of Visual Projection, Candidate's Dissertation, Moscow.
Hagiwara, S., and Morita, H., 1962, J. Neurophysiol., 25(6):721.
Hanbery, J., and Jasper, H., 1953, J. Neurophysiol., 16:252.
Hern, J. E., Landgren, S., Phillips, C. G., and Porter, R., J. Physiol. (London), 161:73.
Hernandez-Peón, R., Guzman-Flores, C., Alcaraz, M., and Fernandez-Guardiola, A., 1956, Fed. Proc., 15:91.

Hernandez-Peón, R., and Donosso, M., 1959, "Influence of attention and suggestion upon subcortical evoked electrical activity in the human brain," in: Proceedings of the International Congress of Neurological Science, Vol. 3, New York, pp. 385-396.

Hess, W. R., 1929, Arch. Psychiat. Nervenkr., 86:287.

Hess, W. R., 1944, Helv. Physiol. Pharmacol. Acta, 2:306.

Hess, W. R., 1949, J. Physiol. (Paris), 41:61a.

Hild, W., Chang, J., and Tasaki, I., 1958, "Electrical responses of astrocytic glia from the mammalian central nervous system cultivated in vitro,'" Experientia, 14(6):220.

Hori, Y., 1965, "Studies on dominant focus of motor cortex including unitary analysis," Abstracts of Papers of the 23rd International Congress of the Physiological Sciences, Tokyo, p. 465.

Hunter, J., and Ingvar, D. H., 1953, "Irradiation pathways for photic metrazol responses in the cat," 3rd International Congress of Electroencephalography and Clinical Neurophysiology, Suppl. III, p. 77.

Hydén, H., 1959, "Biochemical changes in glial cells and nerve cells at varying activity," Proceedings of the 4th International Congress of Biochemistry, Vol. 3, Pergamon Press, London, pp. 64-89.

Hydén, H., 1962, J. Cell Biol., 13:233.

Irwin, D., Knott, J., McAdam, D., and Rebert, C., 1966, Electroenceph. Clin. Neurophysiol., 21(6):538.

Jasper, H., 1949, Electroenceph. Clin. Neurophysiol., 1:405.

Jasper, H., 1960, "Unspecific thalamocortical relations," Handbook of Physiology, Vol. 2, Washington, p. 1307.

Jasper, H., Hunter, J., and Knighton, R., 1948, Trans. Am. Neurol. Ass., 73:210.

Jasper, H., Ricci, G., and Doane, B., 1962, "Patterns of cortical neuronal discharge during conditioned responses in monkeys," in: Electroencephalographic Investigation of Higher Nervous Activity [Russian translation], Izd. AN SSSR, Moscow, pp. 129-146 (G. E. W. Wolstenholme and C. M. O'Connor (Editors): Neurological Basis of Behavior, Ciba Foundation Symposium, London, Churchill, 1958, p. 277).

Kalinin, P. I., 1963, "Changes in cortical bioelectrical activity connected with reinforcement of a cortical dominant focus by afferent stimulation," in: Nervous Mechanisms of Conditioned-Reflex Activity [in Russian], Izd. AN SSSR, Moscow.

Kalinin, P. I., 1965, Effect of Polarization of Specific and Nonspecific Subcortical Structures on a Cortical Dominant Focus, Candidate's Dissertation, Moscow.

Kalinin, P. I., and Liu Han-sheng, 1962, "Role of the mesencephalic and thalamic reticular formations in the formation of a motor dominant focus," Trudy Inst. Vyssh. Nervn. Deyat. Akad. Nauk SSSR, Seriya Fiziol., 7:57.

Kalinin, P. I. and Sokolova, A. A., 1968, "Effect of stimulation of the mesencephalic reticular formation on unit activity in the motor cortex," Zh. Vyssh. Nervn. Deyat., 8(3):528.

Kaplan, I. I., and Ukhtomskii, A. A., 1923, Russk. Fiziol. Zh., 6:71.

Karamyan, A. I., 1959, Fiziol. Zh. SSSR, 45(7):778.

Karamyan, A. I., 1962, Fiziol. Zh. SSSR, 48(7):785.

Karlson, U., and Shultz, R., 1964, Nature, 201:1230.

Kats, K., 1958a, "Nonspecific response in the human EEG under normal conditions and in localized brain lesions," Candidate's Dissertation, Moscow.
Kats, K., 1958b, Zh. Vyssh. Nervn. Deyat., 8:499.
Katsnel'son, R. S., and Vladimirskii, N. D., 1923, Cited by: A. A. Ukhtomskii, The Dominant Focus as a Working Principle of Nerve Centers. Collected Works, Vol. 2, Leningrad University Press, Leningrad, 1950, p. 166.
Kawamura, H., and Sawyer, C. H., 1964, Am. J. Physiol., 207:1379.
Keating, J., and Kempinsky, W., 1960, Electroenceph. Clin. Neurophysiol., 12(4):875.
Kleene, S. K., 1956, "Representation of events in nerve nets and final automata," in: Automata Studies [Russian translation], IL, Moscow, pp. 15-67.
Klosovskii, B. N., 1959, "The construction of the brain," in: Structure and Function of the Reticular Formation and its Place in the System of Analyzers [in Russian], Moscow, pp. 181-198.
Klosovskii, B. N., and Volzhina, N. S., 1956, Vopr. Neirokhir., 1:8.
Knipst, I. N., 1955, "Electrical activity of different levels of the cerebral cortex during defensive conditioning in rabbits," Trudy Inst. Vyssh. Nervn. Deyat AN SSSR. Seriya Fiziol., 1:294.
Kogan, A. B., 1949, Electrophysiological Investigation of the Central Mechanisms of Some Complex Reflexes [in Russian], Izd. AN SSSR, Moscow.
Köhler, W., and O'Connell, D., 1957, J. Cell Comp. Physiol., 49, Suppl. 2:1.
Kononova, E. P., 1926, Anatomy and Physiology of the Occipital Lobes on the Basis of Clinical, Pathological-Anatomical, and Experimental Data [in Russian], Moscow.
Konradi, G. P., 1930, "Conversion of the dominant focus into inhibition," in: Transactions of the Physiological Laboratory, Leningrad State University (Jubilee Collection) [in Russian], Leningrad, pp. 118-132.
Kostyuk, P. G., and Semenyutin, I., 1961, Fiziol. Zh. SSSR, 47:678.
Kotlyar, B. I., and Shul'govskii, V. V., 1966, "Bioelectrical activity of neurons in a dominant focus," in: Electrophysiology of the Central Nervous System [in Russian], Tbilisi, p. 161.
Kozhevnikov, V. A., and Meshcherskii, R. M., 1963, Modern Methods of Analysis of the Electroencephalogram [in Russian], Moscow, Medgiz.
Kuffler, S. W., and Nicholls, J. G., 1965, "How do materials exchange between blood and nerve cells in the brain?" Perspect. Biol. Med., 2(1):69.
Kuffler, S. W., and Nicholls, J. G., 1966, Ergebn. Physiol., 57:1.
Kuffler, S. W., Nicholls, J. G., and Orkand, R. K., 1966, J. Neurophysiol., 29(4):768.
Kunstman, K. I., and Orbeli, L. A., 1924, "Consequences of deafferentation of the hind limb in dogs," Izd. Petrograd. Nauchn. Inst. im. I. G. Lesgafta, 9(2):187.
Kupalov, P. S., 1947, Fiziol. Zh. SSSR, 33(6):699.
Kurepina, M. M., 1959, "The reticular formation of the thalamus in ontogenesis and phylogenesis," in: Structure and Function of the Reticular Formation [in Russian], Moscow, p. 57.
Kuznetsov, S. A., 1963, "Microelectrode investigation of the principal electrophysiological characteristics of cortical single neurons," in: Electrophysiology of the Nervous System [in Russian], Rostov University Press, Rostov-on-Don, pp. 208-209.

Kuznetsova, G. D., 1957, Investigation of the Properties of the Swallowing Dominant and Its Effect on Higher Nervous Activity, Candidate's Dissertation, Moscow.

Kuznetsova, G. D., 1963, "Changes in the cortical steady potential under the influence of the anode of a direct current," in: Nervous Mechanisms of Conditioned-Reflex Activity [in Russian], Izd. AN SSSR, Moscow, pp. 182-195.

Landau, W., Bishop, G., and Clare, M., 1964, J. Neurophysiol., 27(5):788.

Landau, W., Bishop, G., and Clare, M., 1965, J. Neurophysiol., 28(6):1205.

Landgren, S., Phillips, C. G., and Porter, R., 1962a, J. Physiol. (London), 161:91.

Landgren, S., Phillips, C. G., and Porter, R., 1962b, J. Physiol. (London), 161:112.

Larsson, L., 1960, Electroenceph. Clin. Neurophysiol., 12(2):502.

Leyton, A. S., and Sherrington, C. S., 1917, Quart. J. Exp. Physiol., 11:135.

Li, C-L., 1956, J. Physiol. (London), 131:115.

Li, C-L., 1959, J. Neurophysiol., 22(4):436.

Li, C-L., 1962, "Activity of interneurons of the motor cortex," in: Reticular Formation of the Brain. International Symposium, Detroit, USA [Russian translation], Medgiz, Moscow, pp. 405-417.

Li, C-L., and Salmoiraghi, G., 1963, Nature, 198:858.

Libet, B., and Gerard, R., 1941, J. Neurophysiol., 4:438.

Libet, B., and Kahn, I. B., 1947, Fed. Proc., 6:152.

Livanov, M. N., 1962a, "The use of a computer technique to analyze the bioelectrical processes of the brain," in: Biological Aspects of Cybernetics [in Russian], Izd. AN SSSR, Moscow, pp. 112-121.

Livanov, M. N., 1962b, Zh. Vyssh. Nervn. Deyat., 12(3):399.

Livanov, M. N., Korol'kova, T. A., and Frenkel', G. M., 1951, Zh. Vyssh. Nervn. Deyat., 1(4):521.

Lorente de Nó., 1938, in: Fulton, J. F. (Editor), Physiology of the Nervous System, Oxford University Press, New York, p. 229.

Lorente de Nó, R., and Condoris, G. A., 1959, Proc. Nat. Acad. Sci. (Washington), 45:592.

Low, M., Frost, J. D., Borda, R., and Kellaway, P., 1966, Electroenceph. Clin. Neurophysiol., 21:413.

Lur'e, R. N., 1949, After-changes in the Human Electroencephalogram During the Activity of the Visual Analyzer, Candidate's Dissertation, Moscow.

Lur'e, R. N., and Rusinov, V. S., 1955, Probl. Fiziol. Optiki, 11:113.

McAdam, D., Irvin, D., Robert, C., and Knott, I., 1966, Electroenceph. Clin. Neurophysiol., 21:194.

McCulloch, W. S., and Pitts, W., 1956, "Logical calculation of ideas relating to nervous activity," in: Automata [Russian translation], IL, Moscow, pp. 362-384.

Magnitskii, A. N., 1952, "Dominant phenomena in the regulation of the circulation under normal and pathological conditions," in: Nervous Regulation of the Circulation and Respiration [in Russian], Moscow, p. 202.

Magoun, H. W., 1944, Science, 100:549.

Maiorchik, V. E., 1964, Clinical Electrocorticography. Investigations During Neurosurgical Operations [in Russian], Meditsina, Leningrad.

Makarov, P. O., 1939, "Dynamics of excitability, conduction, and the reflex state," Trudy Leningrad Obshch. Estestvoisp., 67:1.

Martin, A. K., and Pilar, G., 1963, J. Physiol. (London), 168:443.
Maruhashi, J., Mezuguchi, K., and Tasaki, G., 1962, J. Physiol., 117:129.
Matsumoto, H., and Ajmone-Marsan, C. A., 1964, Exp. Neurol., 9:286.
Meshcherskii, R. M., 1955, "Changes in electrical activity in the rabbit visual cortex during defensive conditioning," Trudy Inst. Vyssh. Nervn. Deyat. AN SSSR, Seriya Fiziol., 1.
Meshcherskii, R. M., 1966, Corticofugal Effects in the Visual Cortex, Doctoral Dissertation, Moscow.
Meshcherskii, R. M., and Gustson, P. P., 1964, Physiol. Bohemoslov., 13(3):236.
Meyer, A., 1906, Rhythmic Pulsation in Scyphomedusae, Washington, p. 126.
Mingrino, S., Coxe, W., Katz, R., Goldring, S., and O'Leary, J., 1963, Progress in Brain Research, 1:241.
Mishin, L. N., 1963, "Correlation analysis of physiological processes," in: The Application of Mathematical Methods in Biology, No. 2, Leningrad University Press, Leningrad, pp. 94-99.
Mnukhina, R. S., 1961, Zh. Vyssh. Nervn. Deyat., 11(2):346.
Morrell, F., 1961, Ann. New York Acad. Sci., 92:860.
Morrell, F., 1962a, "Microelectrode and steady potential studies suggesting a dendritic locus of closure," in: Electroencephalographic Investigation of Higher Nervous Activity [in Russian], Izd. AN SSSR, Moscow, pp. 54-58. Reprinted in Electroenceph. Clin. Neurophysiol., 13:65, 1960.
Morrell, F., 1962b, Fiziol. Zh. SSSR, 53(3):251.
Morrell, F., 1963, "The action of anodal polarization on cortical single unit activity," in: Higher Nervous Activity [in Russian], Moscow, pp. 52-68.
Moruzzi, G., and Magoun, H., 1949, Electroenceph. Clin. Neurophysiol., 1:455.
Murphy, J. P., and Gellhorn, E., 1945, J. Neurophysiol., 8:341.
Narahashi, T., 1964, J. Cell Comp. Physiol., 64(1):73.
Narikashvili, S. P., 1962, Nonspecific Brain Structures and the Perceptual Function of the Cerebral Cortex [in Russian], Izd. AN Gruz. SSR, Tbilisi.
Naumova, T. S., 1953, Changes in Electrical Activity of the Cortex, Caudate Nucleus, and Medial Geniculate Body During Closure of the Auditory and Motor Systems in Rabbits, Candidate's Dissertation, Moscow.
Naumova, T. S., 1956, Fiziol. Zh. SSSR, 42(4):361.
Neumann, J. von, 1956, "Stochastic logic and synthesis of reliable organisms from unreliable components," in: Automata Studies [Russian translation], IL, Moscow, pp. 68-139.
Novikova, L. A., Rusinov, V. S., and Semiokhina, A. F., 1952, Zh. Vyssh. Nervn. Deyat., 2:844.
Novikova, L. A., and Farber, D. A., 1956, Fiziol. Zh. SSSR, 43(5):341.
Okudzhava, V. M., 1963, "Dendritic Activity in the cerebral cortex," in: Gagra Colloquia, Vol. 4, Izd. AN Gruz. SSR, Tbilisi, pp. 223-337.
O'Leary, J. L., and Goldring, S., 1964, Physiol. Rev., 4(1):91.
Orbeli, L. A., and Kunstman, K. I., 1921, Russk. Fiziol. Zh., 4:253.
Oswald, Z., 1959, Electroenceph. Clin. Neurophysiol., 11(2):341.
Pakovich, B. I., 1960 "Conditions of formation of defensive motor conditioned reflexes," in: Central and Peripheral Mechanisms of Motor Activity of Animals [in Russian], Izd. AN SSSR, Moscow, pp. 86-124.

Pavlov, I. P., 1926, Complete Collected Works [in Russian], Vol. 4, Moscow-Leningrad, 1947.

Pavlov, I. P., 1932, Complete Collected Works [in Russian], Vol. 3, Moscow-Leningrad, 1951.

Pavlygina, R. A., 1956, "Creation of a dominant focus in the hypothalamic region and investigation of its properties," Trudy Inst. Vyssh. Nervn. Deyat. AN SSSR. Seriya Fiziol., 2:124.

Pavlygina, R. A., 1960, "Electrical activity of the cortex, thalamus, hypothalamus, and reticular formation of the brain stem of rabbits during combinations of acoustic and repetitive photic stimulation," Trudy Inst. Vyssh. Nervn. Deyat. AN SSSR. Seriya Fiziol., 5:39.

Pavlygina, R. A., 1962, "Investigation of a focus of excitation of the "single tetanic contraction" type created in the cerebral cortex," Trudy Inst. Vyssh. Nervn. Deyat. AN SSSR. Seriya Fiziol., 7:39.

Pavlygina, R. A., 1964, "Trace phenomena associated with the dominant focus," Proceedings of the 10th Congress of the I. P. Pavlov All-Union Physiological Society [in Russian], Vol. 2, p. 156.

Pavlygina, R. A., and Pozdnyakova, R. A., 1960, "Creation of a dominant focus in the motor cortex by a pulsating direct current," Trudy Inst. Vyssh. Nervn. Deyat. AN SSSR. Seriya Fiziol., 5:49.

Pearlman, A., Goldring, S., and O'Leary, J., 1960, Proc. Soc. Exp. Biol. (New York), 103:600.

Peterson, R. P., and Pepe, F. A., 1961, J. Biophys. Biochem. Cytol., 11:157.

Petrov, F. P., and Lapitskii, D. A., 1926, "Complete restoration of a parabiotic nerve by the anode," Proceedings of the 2nd All-Union Congress of Physiologists [in Russian], Leningrad, p. 295.

Phillips, C. G., 1956, Quart. J. Exp. Physiol., 41:70.

Phillips, C. G., 1959, Quart. J. Exp. Physiol., 44:1.

Phillips, C. G., 1961, "Some properties of pyramidal neurones of the motor cortex," in: The Nature of Sleep, Ciba Foundation, London, pp. 4-24.

Podsosennaya, L. S., 1956, "A dominant focus to diminishing electrodermal stimulation and its effect on defensive conditioned reflexes," Trudy Inst. Vyssh. Nervn. Deyat. AN SSSR. Seriya Fiziol., 2:139.

Pollen, D. A., and Ajmone-Marsan, C. A., 1965, J. Neurophysiol., 28(2):342.

Pressman, Ya. M., 1960, "Interaction between inhibitory and positive conditioned motor reflexes in microintervals of time," in: Central and Peripheral Mechanisms of Motor Activity of Animals [in Russian], Izd. AN SSSR, Moscow, pp. 145-156.

Pribram, K., 1962, "The new neurology: memory, novelty, thought and choice," in: EEG and Behavior, New York, pp. 149-173.

Puchinskaya, L. M., 1960, Byull. Eksperim. Biol. i Med., 50(11):3.

Puchinskaya, L. M., 1961, "Local changes in the human EEG in response to visual afterimages," Byull. Eksperim. Biol. i Med., 52(10):3.

Puchinskaya, L. M., 1963, Local Changes in the Human EEG in Response to Photic Stimulation, Candidate's Dissertation, Moscow.

Puchinskaya, L. M., 1964, Zh. Vyssh. Nervn. Deyat., 14(6):957.
Purpura, D. P., and Grundfest, H., 1956, J. Neurophysiol., 19(6):573.
Purpura, D. P. and McMurtry, J. G., 1965, J. Neurophysiol., 28(1):166.
Rabinovich, M. Ya., 1963, Electrical Responses of Single Layers of the Cortex During Conditioned Reflex Formation, Doctoral Dissertation, Moscow.
Rabinovich, M. Ya., 1967, Zh. Vyssh. Nervn. Deyat., 17(1):3.
Raeva, S. N., 1966, Zh. Vyssh. Nervn. Deyat., 16(1):67.
Ricci, G. F., 1955, Am. J. Physiol., 183:655.
Ricci, G., Doane, B., and Jasper, H., 1957, "Microelectrode studies of conditioning: technique and preliminary results," in: IV Internat. Congr. d'electroencephalographie et de neurophysiologie clinique. Rapports, discussions des sciences neurologiques. Brussels, p. 401.
de Robertis, E., 1958, "Submicroscopic morphology and function of the synapse," in: The Submicroscopic Organization and Function of Nerve Cells, Suppl. 5, pp. 347-369.
Roitbak, A. I., 1953, "An oscillographic study of foci of increased excitability in the cerebral cortex," Trudy Inst. Fiziol. ZN Gruz. SSR, 9:96.
Roitbak, A. I., 1955, Bioelectrical Phenomena in the Cerebral Cortex [in Russian], Tbilisi.
Roitbak, A. I., 1965, "Slow negative cortical surface potentials and inhibition," in: Reflexes of the Brain [in Russian], Nauka, Moscow, pp. 186-196.
Roitbak, A. I., 1970, Neirofiziologiya, 2(4):339.
Rosenblueth, W. A., 1961, "Some quantitative aspects of electrical activity of the central nervous system," in: Current Problems in Biophysics [Russian translation], Vol. 2, IL, Moscow, pp. 282-292.
Rossi, G., and Zanchetti, A., 1960, The Brain-Stem Reticular Formation [Russian translation], IL, Moscow.
Rothballer, A. B., 1956, Electroenceph. Clin. Neurophysiol., 8:602.
Rowland, V., and Goldstone, M., 1963, Electroenceph. Clin. Neurophysiol., 15(3):474.
Rusinov, V. S., 1930a, "Inhibition after excitation in a nerve-muscle preparation," Collected Physiological Materials of Leningrad State University [in Russian], Leningrad University Press, Leningrad, pp. 30-37.
Rusinov, V. S., 1930b, "Differential role of individual parts of a parabiotic segment of nerve in the function of conduction," Trudy Petergof. Estestvennonauchnogo Inst., 7:33.
Rusinov, V. S., 1934, "The relationship between stimulation and excitation during disinhibition of parabiosis by a direct current," Trudy Fiziol. Inst. Leningrad Gos. Univ., 14:10.
Rusinov, V. S., 1947, "The electrophysiological method of studying integrative activity of the nervous system," Proceedings of the 7th All-Union Congress of Physiologists, Biochemists, and Pharmacologists [in Russian], Medgiz, Moscow, pp. 201-204.
Rusinov, V. S., 1949, "The electrophysiological method of studying integrative activity of the nervous system (concluding remarks of the Address to the 7th All-Union Congress of Physiologists)," in: Problems in Soviet Physiology, Biochemistry, and Pharmacology [in Russian], Moscow, pp. 128-131.

Rusinov, V. S., 1951, "Some aspects of the theory of the electroencephalogram and the dominant focus in the cerebral cortex," Proceedings of the 14th Conference on Problems in Higher Nervous Activity [in Russian], Izd. AN SSSR, Moscow—Leningrad, pp. 36-37.

Rusinov, V. S., 1953a, "Electrophysiological analysis of the functions of closure in the cerebral hemispheres in the presence of a dominant focus," Proceedings of the 19th International Physiological Congress [in Russian], Moscow, pp. 147-151.

Rusinov, V. S., 1953b, Vopr. Neirokhir., 17(1):3.

Rusinov, V. S., 1956, "Electrophysiological investigation of the dominant focus at higher levels of the central nervous system," Proceedings of the 20th International Congress of Physiologists [in Russian], Moscow, pp. 350-354.

Rusinov, V. S., 1961, Zh. Vyssh. Nervn. Deyat., 11(5):776.

Rusinov, V. S., 1962, "General and local changes in the electroencephalogram during conditioning," in: Electroencephalographic Investigation of Higher Nervous Activity [in Russian], Izd. AN SSSR, Moscow, pp. 288-296.

Rusinov, V. S., 1965a, "Experimental and clinical investigation of foci of excitation," in: Current Problems in Neurophysiology [in Russian], Moscow—Leningrad, pp. 73-99.

Rusinov, V. S., 1965b, Zh. Vyssh. Nervn. Deyat., 88(2):217.

Rusinov, V. S., and Chugunov, S. A., 1938, Nevropat. i Psikhiat., 7:5.

Ryabinina, M. A., 1959, "A dominant focus produced by pressure in the rabbit cerebral cortex," Trudy Inst. Vyssh. Nervn. Deyat. AN SSSR. Seriya Fiziol., 3:42.

Ryabinina, M. A., 1961, "The role of individual layers of the cerebral cortex in the formation of a dominant focus by means of a direct current," Trudy Inst. Vyssh. Nervn. Deyat. AN SSSR. Seriya Fiziol., 6:211.

Ryabinina, M. A., 1963, "Functional connections of the globus pallidus with the reticular formation and motor cortex as shown by electrophysiological findings," in: Nervous Mechanisms of Conditioned-Reflex Activity [in Russian], Izd. AN SSSR, Moscow, pp. 249-256.

Ryabinina, M. A., 1965, Fiziol. Zh. SSSR, 51(10):1149.

Sager, O., 1961, "A morpho-physiological study of the reticular substance," in: Problems in General Neurophysiology and Higher Nervous Activity [in Russian], First Moscow Medical Institute, Moscow, pp. 129-138.

Samoilov, A. F. (Samoiloff, A.), 1910, Arch. Ges. Physiol., Physiol. Abt., 477.

Samoilov, A. F. (Samoiloff, A.), 1913, Zbl. Physiol., 27(11):575.

Samoilov, A. F. (Samoiloff, A.), 1929, Pflüg, Arch. Ges. Physiol., 222(4):516.

Samoilov, A. F., 1930, "The Circus rhythm of excitation," Selected Papers and Lectures [in Russian], Izd. AN SSSR, Moscow—Leningrad, 1946, pp. 226-260.

Sawa, M., Maruyama, N., and Kaji, S., 1963, Electroenceph. Clin. Neurophysiol., 15(2):209.

Sawa, M., Maruyama, N., Kaji, S., and Nakamura, K., Jap. J. Physiol., 16:126.

Sawa, M., Usuni, K., and Kaji, S., 1965, Electroenceph. Clin. Neurophysiol., 19(3):248.

Sechenov, I. M., 1863, "Investigation of centers maintaining reflex movements in the frog's brain," Selected Works of I. M. Sechenov, I. P. Pavlov, and N. E. Vvedenskii [in Russian], Vol. 3, Medgiz, Moscow, 1952, p. 42.

Sherrington, C. S., 1911, The Integrative Action of the Nervous System, London.
Shkol'nik-Yarros, E. G., 1958, Zh. Vyssh. Nervn. Deyat., 8(1):123.
Shkol'nik-Yarros, E. G., 1965, Neurons and Interneuronal Connections of the Central Visual System [in Russian], Meditsina, Leningrad (English translation, Plenum, New York, 1971).
Shnirman, A. L., 1926, "The conditioned reflex and dominant focus," in: Advances in Reflexology and Physiology of the Nervous System [in Russian], No. 2, Leningrad, pp. 144-158.
Shul'gina, G. I., 1966, "Principles of the systemic organization of cortical neurons during conditioning," in: Electrophysiology of the Central Nervous System [in Russian], Tbilisi, p. 336.
Shul'govskii, V. V., 1967, Unit Activity During the Formation of an Artificial (Polarization) and Conditioned-Reflex Dominant Focus, Candidate's Dissertation, Moscow.
Shumilina, A. I., 1949, Byull. Eksperim. Biol. i Med., 28(12):388.
Shuranova, Zh. P., 1966, Zh. Vyssh. Nervn. Deyat., 16(1):62.
Shvets, T. B., 1958, "Slow electrical processes in the rabbit cerebral cortex," Proceedings of a Conference on Electrophysiology of the Central Nervous System [in Russian], Moscow, p. 138.
Shvets, T. B., 1960a, "Slow changes of potential of the rabbit's cortex under the influence of pressure on the motor area," Trudy Inst. Vyssh. Nervn. Deyat. AN SSSR. Seriya Fiziol., 4:115.
Shvets, T. B., 1960b, "Slow electrical processes in the rabbit's cortex during closure of a temporary connection," Trudy Inst. Vyssh. Nervn. Deyat. AN SSSR. Seriya Fiziol., 5:58.
Shvets, T. B., 1963, Investigation of the Steady Cortical Potential Level in Rabbits During Formation of a Dominant Focus and Defensive Conditioning, Candidate's Dissertation, Moscow.
Sloan, N., and Jasper, H., 1950, Electroenceph. Clin. Neurophysiol., 2:317.
Smirnov, G. D., 1956, Uspekhi Sovr. Biol., 10(3/6):320.
Smirnov, G. D., 1968, Vestn. Akad. Nauk SSSR, 1:48.
Sokolov, E. N., 1965, "Inhibition in analyzer activity," in: Reflexes of the Brain [in Russian], Nauka, Moscow, pp. 72-81.
Sokolova, A. A., 1954, Electrical Activity of the Cortex and Subcortical Structures of the Rabbit in the Presence of a Cortical Dominant Focus, Candidate's Dissertation, Moscow.
Sokolova, A. A., 1958, Zh. Vyssh. Nervn. Deyat., 8(4):593.
Sokolova, A. A., 1959, Zh. Vyssh. Nervn. Deyat., 9(5):759.
Sokolova, A. A., 1965, Zh. Vyssh. Nervn. Deyat., 15(5):878.
Sokolova, A. A., and Khon Sek Bu, 1957, Zh. Vyssh. Nervn. Deyat., 7(1):135.
Sokolova, A. A., and Lipenetskaya, T. D., 1966, Zh. Vyssh. Nervn. Deyat., 16(6):1055.
Storm van Leeuwen, W., 1961, "Computer techniques in EEG analysis," Electroenceph. Clin. Neurophysiol., Suppl. 20:37.
Stoupel, N., and Terzuolo, C., 1954, Acta Neurol. Belg., 54:239.
Strumwasser, F., and Rosenthal, S., 1960, Am. J. Physiol., 198(2):405.

Tabushi, K., 1965, Cortical dc Potential Changes Associated with Spontaneous Sleep in Cats, Abstracts of Proceedings of the 23rd International Physiological Congress, Tokyo, p. 450.

Tasaki, I., 1959, J. Physiol. (London), 148:306.

Tasaki, I., and Chang, H., 1958, Science, 128:1209.

Terzuolo, C. A., and Bullock, T. H., 1956, Proc. Nat. Acad. Sci. (Washington), 42(9):687.

Tobias, J. M., 1959, Ann. Rev. Physiol., 21:299.

Tsypkin, Ya. Z., 1965, Adaptation, Learning, and Self-Teaching in Automatic Systems (Results, Problems, Perspectives) [in Russian], Moscow.

Udel'nov, M. G., 1938, Fiziol. Zh. SSSR, 25:5.

Uflyand, Yu. M., 1924, Fiziol. Zh. SSSR, 7(1-6):167.

Uflyand, Yu. M., 1925, "The natural dominant focus in the male frog during the embracing reflex," in: Advances in Reflexology and the Physiology of the Nervous System [in Russian], No. 1, Leningrad–Moscow, pp. 80-94.

Uflyand, Yu. M., 1927, Fiziol. Zh. SSSR, 10(5):363.

Ukhtomskii, A. A., 1911, "Dependence of cortical motor effects on extraneous central effects," Collected Works [in Russian], Vol. 1, Leningrad University Press, Leningrad, 1950, pp. 31-162.

Ukhtomskii, A. A., 1923, "The dominant focus as a working principle of neural centers," Collected Works [in Russian], Vol. 1, Leningrad University Press, Leningrad, 1950, pp. 163-172.

Ukhtomskii, A. A., 1924, "The dominant focus and integral image," Collected Works [in Russian], Vol. 1, Leningrad University Press, Leningrad, 1950, p. 192.

Ukhtomskii, A. A., 1925, "The principle of the dominant focus," Collected Works [in Russian], Vol. 1, Leningrad University Press, Leningrad, 1950, pp. 197-201.

Ukhtomskii, A. A., 1926, "The state of excitation in the dominant focus," Collected Works [in Russian], Vol. 1, Leningrad University Press, Leningrad, 1950, pp. 208-220.

Ukhtomskii, A. A., 1932, "Problems in stationary excitation," Collected Works [in Russian], Vol. 2, Leningrad University Press, Leningrad, 1951, pp. 54-61.

Ukhtomskii, A. A., 1937, "The present position of N. E. Vvedenskii's school," Collected Works [in Russian], Vol. 2, Leningrad University Press, Leningrad, 1951, p. 112.

Ukhtomskii, A. A., and Vinogradov, M. M., 1925, "Inertia of the dominant focus," in: Ukhtomskii, A. A., Collected Works [in Russian], Vol. 1, Leningrad University Press, Leningrad, 1950, pp. 202-207.

Vanasupa, P., Goldring, S., O'Leary, J., and Winter, L., 1959, J. Neurophysiol., 22(3):273.

Vasil'ev, L. L., 1924, "The fundamental functional state of the nervous system," in: Advances in Reflexology and the Physiology of the Nervous System [in Russian], Vol. 1, Leningrad–Moscow, pp. 1-41.

Vasil'eva, V. M., 1965, Cited by Rusinov, V. S., "Experimental and clinical investigation of foci of excitation," in: Current Problems in Neurophysiology [in Russian], Izd. AN SSSR, Moscow–Leningrad, p. 80.

BIBLIOGRAPHY

Vasil'eva, V. M., Nezlina, N. I., and Ivannikova, T. N., 1966, Zh. Vyssh. Nervn. Deyat., 16(6):1119.
Vasilevskii, N. N., 1965, Zh. Vyssh. Nervn. Deyat., 15(3):529.
Vasilevskii, N. N., 1966, "Neuronal mechanisms of formation of temporary connections at the cortical level," in: Electrophysiology of the Central Nervous System [in Russian], Tbilisi, pp. 63-64.
Verzilova, O. V., 1966, "The mechanism of formation of the flexor dominant focus," in: Problems in Neurophysiology [in Russian], Meditsina, Moscow, pp. 125-135.
Vetokhin, I. A., 1929, Russk. Fiziol. Zh., 12(2):85.
Vetyukov, I. A., 1930, "The importance of frequency of electrical stimulation for the production of a spinal dominant focus," in: Transactions of the Physiological Laboratory of Leningrad State University (Jubilee Collection) [in Russian], Leningrad, p. 145.
Vinogradov, M. I., 1914-1915, "The distorted effect of electrotonus on excitability and conduction of a nerve in parabiosis," Raboty Sankt-Peterburg. Univ., (9-10):145.
Vinogradov, M. I., 1923, Russk. Fiziol. Zh., 6.
Vinogradov, M. I., 1925, "The dominant focus and reflex distortion," in: Advances in Reflexology and the Physiology of the Nervous System [in Russian], Vol. 1, Leningrad–Moscow, pp. 66-79.
Vinogradov, M. I., 1929, Russk. Fiziol. Zh. 6(2):3.
Vinogradov, M. I., 1930, "The history of the study of the dominant focus," Collection in Honour of 25 Years of A. A. Ukhtomskii's Activity [in Russian], Leningrad University Press, Leningrad, pp. 5-18.
Vinogradov, M. I., and Konradi, G. P., 1928, Med.-Biol. Zh., 2:63.
Voronin, L. G., 1952, Analysis and Synthesis of Complex Stimuli of Higher Animals [in Russian], Leningrad, p. 193.
Voronin, L. G., 1957, Lectures on the Comparative Physiology of Higher Nervous Activity [in Russian], Moscow University Press, Moscow.
Voronin, L. L., 1966, Some Problems in the Functional Organization of Neurons of the Motor Cortex of the Waking Rabbit (Microelectrode Investigation), Candidate's Dissertation, Moscow.
Voronin, L. L., 1970a, Neirofiziologiya, 2(5):460.
Voronin, L. L., 1970b, Neirofiziologiya, 2(6):601.
Voronin, L. L., 1971, Neirofiziologiya, 3(1):3.
Voronin, L. L., and Skrebitskii, V. G., 1964, "Changes in single cortical unit reactivity under the influence of polarization and against the background of arousal reactions," Proceedings of the 10th Congress of the All-Union Physiological Society [in Russian], Vol. 2, No. 1, Nauka, Moscow, p. 179.
Voronin, L. L., and Skrebitskii, V. G., 1965, Byull. Eksperim. Biol. i Med., 59(5):3.
Voronin, L. L., Skrebitskii, V. G., and Sharonova, I. N., 1971, Uspekhi Fiziol. Nauk, 2(1):116.
Voronin, L. L., and Solntseva, E. I., 1969, Zh. Vyssh. Nervn. Deyat., 19(5):828 (Neurosci. Transl. 13:65-74, 1970).
Vorontsov, D. S., 1925, Pflüg. Arch. ges. Physiol., 210:672.

Vvedenskii, N. E., 1881, Complete Collected Works, Vol. 1, Leningrad University Press, Leningrad, 1951, pp. 126-144.
Vvedenskii, N. E., 1899, Complete Collected Works [in Russian], Vol. 6, Leningrad University Press, Leningrad, 1956, pp. 184-188.
Vvedenskii, N. E., 1901-1903, "Excitation, inhibition, and narcosis," Complete Collected Works [in Russian], Vol. 4, Leningrad University Press, Leningrad, 1951, pp. 9-146.
Vvedenskii, N. E., 1906, "Excitation and inhibition in the reflex apparatus during strychnine poisoning," Complete Collected Works [in Russian], Vol. 4, Leningrad University Press, Leningrad, 1951, pp. 202-269.
Vvedenskii, N. E., and Ukhtomskii, A. A., 1908, Cited by Vvedenskii, N. E., Complete Collected Works [in Russian], Vol. 4, Leningrad University Press, Leningrad, 1953, p. 318.
Walter, W. G., 1950, J. Ment. Sci., 96:1.
Walter, W. G., 1964, Electroenceph. Clin. Neurophysiol., 16(6):623.
Walter, W. G., 1965, "Contingent negative variation as an electrocortical sign of sensorimotor reflex association in man," in: Reflexes of the Brain [in Russian], Nauka, Moscow, pp. 365-381.
Walter, V. J., and Walter, W. G., 1949, Electroenceph. Clin. Neurophysiol., 1(1):57.
Washizu, Y., 1965, Comp. Biochem. Physiol., 15(4):535.
Wenckebach, K. F., and Winterberg, 1927, Die Unregelmässige Herztätigkeit.
Whitear, M., 1965, Phil. Trans. Roy. Soc. B., 248:437.
Wiener, N., 1961, Nonlinear Problems in the Theory of Random Processes [Russian translation], Moscow.
Wurtz, R. H., 1964, Electroenceph. Clin. Neurophysiol., 18(7):649.
Zhegalkina, N. G., 1967, Investigation of the Rabbit EEG by Correlation Analysis During Polarization of the Sensorimotor Cortex, Candidate's Dissertation, Moscow.
Zhukova, G. P., 1959, Arkh. Anat. Gistol. i Émbriol., 30:32.

QP
376
R8713
1973

DEC 17 1973